ARAPAHO & ROOSEVELT NATIONAL FORESTS

Roosevelt National Forest was originally part of the Medicine Bow Forest Reserve established in 1897, it became the Colorado National Forest in 1910. In 1932, President Herbert Hoover renamed it in honor of the person most responsible for creating the National Forest System, President Theodore Roosevelt.

Arapaho National Forest established July 1, 1908, by President Theodore Roosevelt. This forest is named after the plains Indian tribe which frequented the area for summer hunting. Management of the Districts of the Arapaho National Forest is now divided among three national forests. Boulder, Sulphur and Clear Creek Districts are administered as part of the Arapaho and Roosevelt National Forests, Dillon District by White River National Forest.

The headquarters of the Arapaho and Roosevelt National Forests and Pawnee National Grassland is in Fort Collins. The three units, which were administratively combined in 1973, include nearly 1.5 million acres, about 20 percent of the area is in the National Wilderness Preservation System, in nine wildernesses.

Because of their nearness to the high populations of Front Range communities, and area of nationally known significance, the Arapaho and Roosevelt rank among the top national forests for year-round recreation use. They are a major part of the scenic backdrop for tourism in Colorado. Camping, hiking, fishing, hunting, skiing and other winter sports, and driving for pleasure are popular activities.

In addition to the major summer and winter recreation programs, the forests and grassland are managed for a variety of other public benefits under the Multiple Use-Sustained Yield Act of 1960, and the National Forest Management Act of 1976. These public lands are a vital source of water, wildlife and fish habitat, wood, forage for grazing, and minerals and energy.

CULTURAL HISTORY

The Arapaho and Roosevelt National Forests are rich in cultural history. The first human use probably occurred between 5,000 and 10,000 years B.C. Native tribes occupied the forests and plains up through the early 19th century. Then early day ranching, mining, and agriculture uses developed on the lands. Signs of these early uses are old mines, pack trails, wagon roads, narrow and standard gauge railroad routes, and prospector, trapper, and homestead cabins. If you discover any remnants and relics of our past, leave them in place for the next visitor to appreciate. Cultural artifacts may not be collected from federal lands, please report them to the district office.

WILDERNESS

Nine wildernesses are managed to protect their natural ecosystems. They also offer opportunities for human isolation, solitude, self-reliance, and challenge while hiking cross-country or on trails. Please practice low impact camping techniques and follow local regulations for use of these areas.

ARAPAHO NATIONAL RECREATION AREA (ANRA)

The ANRA covers over 36,000 acres and contains five major lakes, often referred as "the Great Lakes of Colorado."

Boating and fishing are primary activities, with many developed campgrounds, picnic areas, and hiking trails available for public recreation and enjoyment.

DISPERSED RECREATION

Although off-road motorized travel is restricted on the Arapaho and Roosevelt National Forests, you can enjoy motorized travel challenges on unimproved roads open to primitive sites for hunting, fishing, camping, picnicking, hiking, horseback riding, and viewing wildlife and scenery. Many roads are passable only by high-clearance, and/or four wheel drive vehicles. Within the national forest boundary are many parcels of privately owned land, please respect the rights of private landowners. Be especially careful with anything that could cause a fire in the backcountry, make sure any campfires and smoking materials are completely put out.

ROCKY MOUNTAIN NATIONAL PARK

Included in the guide are 30 trails within Rocky Mountain National Park. A feature of the trails is the marked difference found with the changing elevations. At lower levels montane life zone; pines, aspens, meadows, wildflowers and glades. Above 9,000 feet; Engleman spruce and subalpine fir. Openings in these cool dark forests produce wildflowers gardens of luxuriance and beauty. Then the trees disappear and you are in alpine tundra, a harsh fragile world. Here, more than one-quarter of the plants you see can be found in the Arctic. The park encompasses many worlds. We invite you to explore them.

Maps and Descriptions of 86 Trails!
Trails for the Entire Family
Arapaho/Roosevelt National Forests
and
Rocky Mountain National Park Trails

Trail Information
- Trail Name • Number
- Difficulty • Access
- Distance • Attractions
- Narrative • Activities
- Use • Topographic Map Name
- Beginning and Ending Elevations
- Regulations & Phone numbers

Recreation Information
- Hiking
- Mountain Biking
- Horses
- ATV's
- Motorcycles
- Campground locations

HOW TO USE THIS GUIDE

In this guide Arapaho/Roosevelt National Forests are divided into 23 map areas. The numbered map located in front of area locates trails, campgrounds, rivers, lakes, wilderness boundaries and other physical features. Map scale is approximately 1"= 2 miles. Trails and campgrounds are numbered on the map with a symbol. The campground information is located in each area and referenced to the map and campground map number.

Trail "quick reference" information bar heading is provided for each trail.
BAR HEADING EXPLANATION

Forest Service Trail number. | Trail Name | ONE-WAY distance of trail. | Beginning and ending trail elevations. | Ranger District that manages trail

Trail No.	Trail Name	Map Loc.	Distance	Difficulty	Beginning Elev.	Ending Elev.	Ranger District
712	**Cottonwood**	**J 7**	**1.6 mi**	**Moderate**	**10,800'**	**10,600'**	**Canyon Lakes**

Map coordinates of trail location on Arapaho/Roosevelt National Forest map.

Forest Service trail rating.

TEXT EXPLANATION

ACCESS: Directions to reach trailhead.

ATTRACTIONS: Describes scenery, destination, and wildlife that one might encounter.

NARRATIVE: Describes terrain, use and conditions of trail.

USE: How many people use the trail. Categories: Heavy, medium and light.

ACTIVITIES: Suggested recreation activities. NOTE: Trails may or may not be restricted to activities shown - call first.

USGS: Name of U.S. Geological Survey 7 1/2' topographic quadrangle on which the trail is shown. Scale 1" = 2000 feet.

NOTE: Information contained in this guide is for general reference only. Use good judgement when planning your trip. Forest Service difficulty trail ratings and directions are general guides, physical condition, age, weather, altitude and experience are factors that should be considered when planning your trip. If you have questions about conditions or trail routes contact the local District Office. The intended use of this guide is for trip planning only. Outdoor Books & Maps, Inc. is not responsible for mishap or injury from other than its intended use.

RANGER DISTRICT INFORMATION

Boulder Ranger District . . . .(303) 444-6600
2995 Baseline Road, Room 110
Boulder, CO 80303

Clear Creek Ranger District . . . .(303) 567-2901
101 Chicago Creek, PO Box 3307
Idaho Springs, CO 80452

Canyon Lakes Ranger District . . . .(970) 498-2770
1311 South College Ave.
Fort Collins, CO 80524

Pawnee National Grass Lands Ranger Dist. (970) 353-5004
660 "O" Street
Greeley, CO 80631

Sulphur Ranger District . . . . (970) 887-4100
9 Ten Mile Drive, PO Box 10
Granby, CO 80446

Forest Supervisor Office . . . . (970) 498-1100
240 West Prospect Road
Fort Collins, CO 80526

MAP SOURCES

Arapaho/Roosevelt Forest Service Maps are sold at Forest Service offices, sporting goods and map stores. All you need for a trouble free trip is a Forest Service map, this Guide, and a U.S. Geologic Topographic map showing the trail. This combination of information and maps is unbeatable!

METHOD FOR RATING TRAIL DIFFICULTY

Four categories for degree of difficulty are as follows:

EASY:
A. Route is mostly level with short uphill/downhill sections.
B. Excellent to good tread surface and clearance.
C. Absence of navigational difficulties/hazards.

MODERATE:
A. Route is level to sloping with longer uphill/downhill sections.
B. Good-to-fair surface and clearance.
C. Minimal navigational difficulties/hazards.

MORE DIFFICULT:
A. Route is level to steep with sustained uphill/downhill sections.
B. Fair to poor surface and clearance.
C. Short sections involving significant navigational difficulties/hazards.

MOST DIFFICULT:
A. Route is mostly steep with sustained uphill/downhill sections.
B. Poor-to-nonexistent tread surface and clearance.
C. Longer sections involving significant navigational difficulties/hazards.

Any rating (i.e., Moderate trail difficulty) found in this book is based on the above scale which has been established for Forest Service purposes. This difficulty rating system is a general guide to help evaluate the trail. Know your limitations!

NOTE: Information contained in this guide is for general reference and trip planning only. Use good judgement! When planning a trip physical condition, age, altitude and weather conditions should be considered. If you have questions about conditions or trail routes contact the local Forest Ranger. Outdoor Books is not responsible for any injury or mishap resulting in the use of this guide for other than its intended use for trip planning.

East Rainbow Lake -- W. Parsons, USFS

Acknowledgements
U.S. Forest Service
Colorado Division of Parks & Outdoor Recreation
U.S. Geological Survey
National Parks Service
Staff
Kristin Alexander
Dody Olofson
Cindi Alexander
Jackie Quintana
Graphics & Cover Design
Grasman Design
Rocky Mountain National Park
Don and Roberta Lowe
Editor & publisher
Jack O. Olofson

Outdoor Books & Maps, Inc.
11270 County Road 49   Hudson, CO 80642
Phone: (303) 536-4640, (800) 952-5342, FAX (303) 536-4641
E Mail: obm@iguana.ruralnet.net
ISBN 0-930657-08-X
© Copyright 1996 Outdoor Books & Maps, Inc. Hudson, CO
Revised 1998

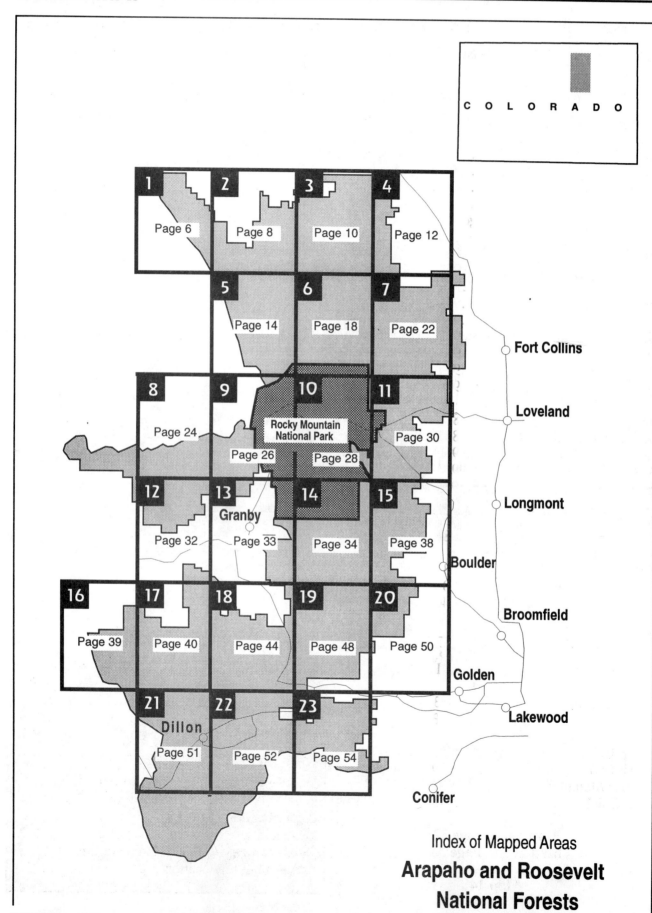

COLORADO

| 1 Page 6 | 2 Page 8 | 3 Page 10 | 4 Page 12 |

| 5 Page 14 | 6 Page 18 | 7 Page 22 |

Fort Collins

| 8 Page 24 | 9 | 10 Rocky Mountain National Park | 11 Page 30 |

Page 26   Page 28

Loveland

| 12 Page 32 | 13 Granby Page 33 | 14 Page 34 | 15 Page 38 |

Longmont

Boulder

| 16 Page 39 | 17 Page 40 | 18 Page 44 | 19 Page 48 | 20 Page 50 |

Broomfield

Golden

| 21 Dillon Page 51 | 22 Page 52 | 23 Page 54 |

Lakewood

Conifer

Index of Mapped Areas
Arapaho and Roosevelt
National Forests

Rocky Mountain National Park Trails Begin on Page 57.
Rocky Mountain National Park Table of Contents Page 59.

MAP 1

No trail descriptions for this map

No campgrounds located on this map

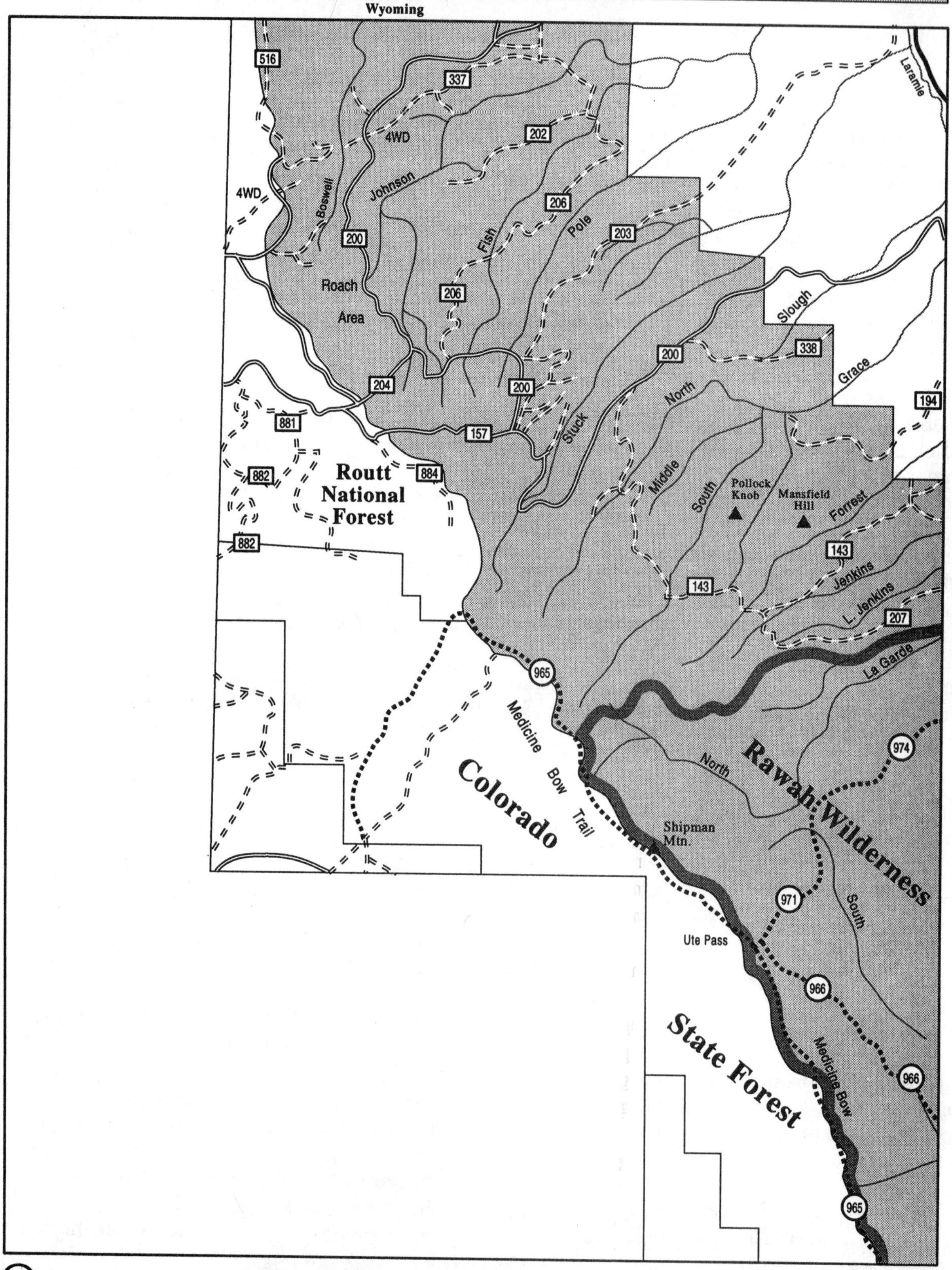

Wyoming

516

337

202

4WD

4WD

Boswell

Johnson

200

Roach

Area

206

Fish

Pole

206

203

204

200

157

Stuck

North

200

338

Grace

194

881

Slough

882

Routt
National
Forest

884

Middle

South

Pollock
Knob

Mansfield
Hill

Forrest

882

143

Jenkins

143

L. Jenkins

207

La Garde

965

Medicine

Bow

Trail

Colorado

North

Rawah Wilderness

974

Shipman
Mtn.

971

South

Ute Pass

966

State Forest

Medicine Bow

966

965

Ties to Map 2 Page 8

⑪⓪ Trail Number Symbol   ❷△ Campground Symbol & Number

Map 1

Trail Name	Trail Number	Activities
Beartracks	43	Hiking, Horses, Fishing
Beaver Creek	942	Hiking, Horses, Fishing
Beaver Crk-Mitchell Lake-Mt Audubon	911/912/913	Hiking
Beavers Meadows	44	Hiking, Horses, Fishing
Big South	944	Hiking, Biking, Fishing
Blue Lake	959	Hiking, Horses
Bob, Betty & King Lakes	810/901	Hiking, Horses, Fishing
Bottle Pass	16	Hiking, Horses
Browns Lake	941	Hiking, Horses, Fishing
Chief Mountain	58	Hiking, Horses, Biking
Crosier Mountain	931	Hiking, Horses, Biking
Darling Creek	18	Hiking, Horses. Fishing
Flowers	939	Hiking, Horses
Foothills Nature-Round Mtn	831	Hiking, Horses
Grays Peak	54	Hiking, Horses, Biking
Greyrock Nat. Recreation	946	Hiking, Horses, Biking
Hourglass	984	Hiking, Horses, Fishing
Killpecker	956	Hiking, Horses, Biking, ATV"s, Motorcycles
Kinney Creek	22	Hiking, Horses, Fishing
Lake Evelyn	15	Hiking, Fishing
Lily Mountain	933	Hiking, Horses
Link	963	Hiking, Horses, Fishing
Lion Gulch	949	Hiking, Horses, Fishing
Little Beaver-Fish Creek	855/1009	Hiking, Horses, Biking
Mcintyre Creek	966	Hiking, Horses, Fishing
Medicine Bow	965	Hiking
Mirror Lake	943	Hiking, Horses, Fishing
Molly Lake-Elkhorn Cr.	174	Hiking, Horses
Montgomery Pass	986	Hiking, Horses, Biking
Mount Margeret	167	Hiking, Biking
Mount McConnel Nat. Rec.	936	Hiking
Mummy Pass-Emmaline Lake	854	Hiking, Horses, Fishing
Niwot Ridge-Jean Luning	906/907	Hiking
North Fork	929	Hiking, Horses
North Lone Pine	953	Hiking, Horses, Biking
Parkview Mountain	-----	Hiking, Horses, Biking
Pawnee Pass-Isabell Glacier	907/908	Hiking
Ptarmigan Pass	35	Hiking
Rainbow Lakes	918	Hiking, Horses, Fishing
Rawah Lakes	960/961	Hiking, Horses
Red Deer Cutoff	824/910/914	Hiking, Horses, Fishing
Resthouse Meadows	57	Hiking, Horses

Trail Name	Trail Number	Activities
Roaring Creek	952	Hiking, Horses, Biking
Silver Dollar Lake	79	Hiking, Horses, Biking Fishing
South Fork Loop	21	Hiking, Horses, Biking, Fish
St Louis Divide-Jones Pass	17	Hiking, Horses, Fishing
St Vrain Mtn-St Vrain Glacier	915/917	Hiking, Horses
Trap Park	866	Hiking, Horses
Ute-James Peak Lake	803/804	Hiking, Horses Biking, Fishing
Ute Pass	31	Hiking, Horses
Ute Peak	24	Hiking, Horses, Fishing
West Branch	960	Hiking, Horses, Fishing
Williams Peak	25	Hiking, Horses
Young Gulch	837	Hiking, Horses, Biking
Zimmerman	940	Hiking, Horses
Zimmerman Lake	977	Hiking, Horses Biking, Fishing

Rocky Mountain National Park

Trail Name	Trail Number	Activities
Andrews Tarn	17	Hiking, Horses
Bierstadt lake	13	Hiking, Horses
Black Lake	19	Hiking, Horses
Bluebird Lake	29	Hiking, Horses
Chasm Lake	23	Hiking, Horses
Crater	5	Hiking, Horses
Cub Lake & Mill Crk Basin	10	Hiking, Horses
Deer Mountain	9	Hiking, Horses
East Inlet	24	Hiking, Horses
Emerald Lake	15	Hiking, Horses
Eugenia Mine	21	Hiking, Horses
Fern & Odessa Lakes	11	Hiking, Horses
Finch Lake	30	Hiking, Horses
Flattop Mountain	14	Hiking, Horses
Gem Lake	8	Hiking, Horses
La Poudre Pass	4	Hiking, Horses
Lake Hayiyaha	16	Hiking, Horses
Lake Helene	12	Hiking, Horses
Lake of the Clouds	2	Hiking, Horses
Lawn & Crystal Lakes	7	Hiking, Horses
Lion Lake	27	Hiking, Horses
North Inlet	1	Hiking, Horses
North Longs Peak	20	Hiking, Horses
Sand Beach Lake	26	Hiking, Horses
Shadow Mountain Lookout	25	Hiking, Horses
Sky Pond	18	Hiking, Horses
Thunder Lake	28	Hiking, Horses
Thunder Pass	3	Hiking, Horses
Twin Sisters Peak	22	Hiking, Horses
Ypsilon Lake	6	Hiking, Horses

MAP 2

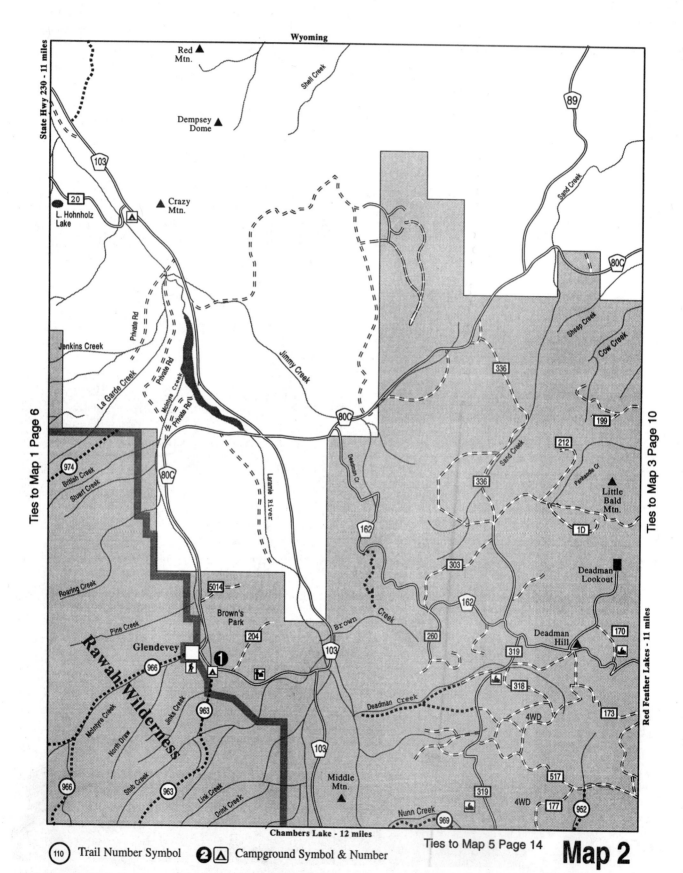

Map 2

Wyoming

State Hwy 230 - 11 miles

Ties to Map 1 Page 6

Ties to Map 3 Page 10

Red Feather Lakes - 11 miles

Red Mtn.

Shell Creek

Dempsey Dome

89

103

20

L. Hohnholz Lake

Crazy Mtn.

Sand Creek

80C

Jenkins Creek

Private Rd

Jimmy Creek

Sheep Creek

Cow Creek

La Garde Creek

Private Rd

McIntyre Creek

336

199

Private Rd

80C

212

Little Bald Mtn.

974

British Creek

Stueri Creek

80C

Laramie River

Deadman Cr

336

Panhandle Cr

1D

162

Deadman Lookout

303

Roaring Creek

5014

162

Pine Creek

Brown's Park

Brown

Creek

260

Deadman Hill

319

170

Glendevey

204

103

318

Rawah Wilderness

966

963

Jinks Creek

Deadman Creek

173

4WD

McIntyre Creek

North Draw

Stub Creek

Link Creek

Drink Creek

103

Middle Mtn.

517

966

963

319

4WD

177

952

Nunn Creek

969

Chambers Lake - 12 miles

Ties to Map 5 Page 14

(110) Trail Number Symbol   ❷⛺ Campground Symbol & Number

MAP 2

963 Trail Number	Trail Name	Map Loc.	Distance	Difficulty	Beginning Elev.	Ending Elev.	Ranger District
	Link	D 2	9.0 mi	Moderate	8,400'	10,600'	Canyon Lakes

ACCESS: 9 miles north of Fort Collins on Highway 287. Continue 59 miles west on Highway 14 and then 16 miles north on County Road 103, Laramie River Road, 2 miles east on County Road 190, Glendevey Road. **ATTRACTIONS:** The trail climbs through a lodgepole pine forest to the Big McIntyre Burn, where excellent views of the Laramie River Valley, north to Wyoming and south to the Poudre Canyon are available. The Link Trail intersects the Medicine Bow Trail, the McIntyre Lake Trail, and finally joins the Rawah Trail at Rawah Lake #1. **USE:** Moderate. **ACTIVITIES:** HIKING, HORSES, FISHING. **USGS:** RAWAH LAKES, GLENDEVEY QUADS. **MAPS:** 2 and 5

966 Trail Number	Trail Name	Map Loc.	Distance	Difficulty	Beginning Elev.	Ending Elev.	Ranger District
	McIntyre Creek	D 2	8.8 mi	Easy	8,400'	10,800'	Canyon Lakes

ACCESS: From Fort Collins, go 9 miles north on Highway 287. Continue 59 miles west on Highway 14 and then go 15 miles north on County Road 103, Laramie River Road. Finally drive 3 miles west on County Road 190, Glendevey Road. **ATTRACTIONS:** This trail follows McIntyre Creek west to Housmer Park, then veers south still following the creek to the McIntyre beaver ponds. The portion of the trail between Housmer Park and the beaver ponds is not well maintained and may be boggy in many places. The ponds provide excellent trout fishing. **USE:** Low. **ACTIVITIES:** HIKING, HORSES, FISHING. **USGS:** RAWAH LAKES, GLENDEVEY, SHIPMAN MTN. QUADS. **MAPS:** 1, 2 and 5.

CAMPGROUNDS LOCATED IN MAP 2

Map No.	Name	Fee	No. of Units	Max. Length	Elev.	Toilets	Water	Ranger District
1.	Browns Park	$	28	30'	8,400'	Yes	No	Canyon Lakes

Clear Creek -- G. Lloyd USFS

MAP 3

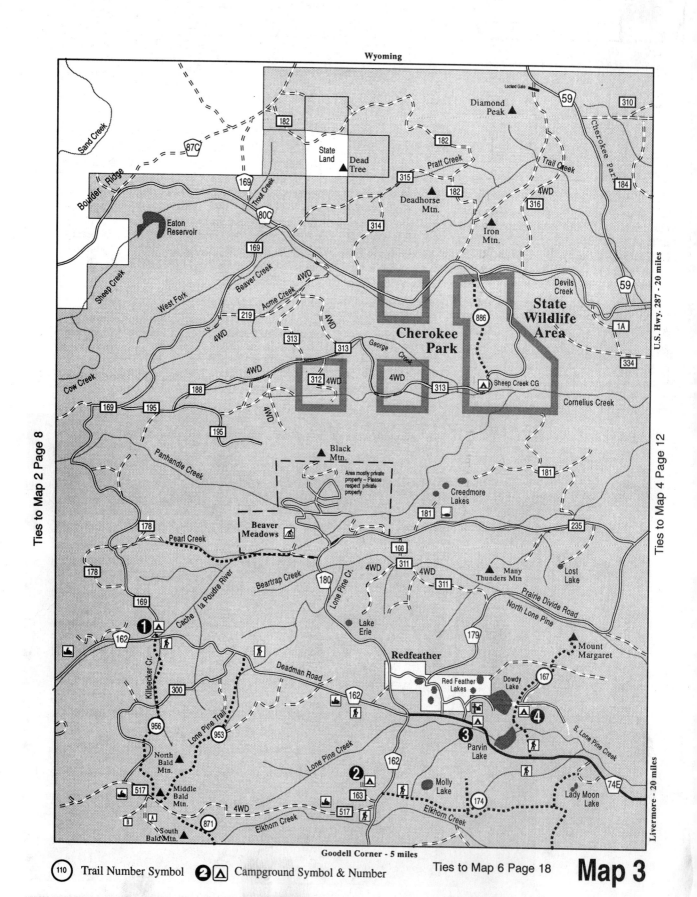

Wyoming

Sand Creek

Boulder Ridge

182

87C

169

State Land

Dead Tree

Diamond Peak ▲

Locked Gate

59

310

Pratt Creek

315

Trail Creek

Cherokee Par...

184

Deadhorse Mtn.

182

4WD

316

59

Trout Creek

80C

314

Iron Mtn. ▲

Eaton Reservoir

169

Devils Creek

1A

169

Sheep Creek

West Fork

Beaver Creek

Acme Creek

4WD

4WD

219

4WD

313

Cherokee Park

886

State Wildlife Area

313

Cow Creek

4WD

312 4WD

George Creek

4WD

313

Sheep Creek CG ▲

Cornelius Creek

188

169   195

195

Panhandle Creek

Black Mtn. ▲

Area mostly private property – Please respect private property

181

Creedmore Lakes

235

178

Pearl Creek

Beaver Meadows

181

178

169

Beartrap Creek

180

Lone Pine Cr.

4WD

311

4WD

311

Many Thunders Mtn ▲

Lost Lake

Cache la Poudre River

Lake Erie

106

Prairie Divide Road

North Lone Pine

162

① △ 🚶

Kilpecker Cr.

300

Deadman Road

162

Redfeather

179

Mount Margaret ▲

956

Lone Pine Trail

953

Red Feather Lakes

Dowdy Lake

167

④ △

North Bald Mtn. ▲

Lone Pine Creek

162

③

Parvin Lake

S. Lone Pine Creek

517

Middle Bald Mtn. ▲

② △

163

Molly Lake

174

74E

B

South Bald Mtn. ▲

871

4WD

517 🚶

Elkhorn Creek

Elkhorn Creek

Lady Moon Lake

Ties to Map 2 Page 8

Ties to Map 4 Page 12

U.S. Hwy. 287 - 20 miles

Livermore - 20 miles

Goodell Corner - 5 miles

(110) Trail Number Symbol   ② △ Campground Symbol & Number

Ties to Map 6 Page 18

Map 3

MAP 3

167	Trail Name	Map Loc.	Distance	Difficulty	Beginning Elev.	Ending Elev.	Ranger District
Trail Number	**Mount Margeret**	G 2	3.9 mi	Easy	8,090'	7,960'	Canyon Lakes

ACCESS: 20 miles north on Highway 287 from Fort Collins, 21 miles west on County Road 74E, Red Feather Lakes Road. **ATTRACTIONS:** The trail crosses South Lone Pine Creek and continues through a lodgepole pine forest that was once logged to treat heavy infestation of pine beetle. It cuts into an old logging road which can be taken, by hikers and bicyclists only to Dowdy Lake Campground. The trail passes a pond and stops at the base of Mount Margaret. **USE:** Low. **ACTIVITIES:** HIKING, MTN BIKING. **USGS:** RED FEATHER LAKES QUAD. **MAP:** 3.

174	Trail Name	Map Loc.	Distance	Difficulty	Beginning Elev.	Ending Elev.	Ranger District
Trail Number	**Molly Lake - Elkhorn Cr**	G 2	1.5-2.0 mi	Easy	8,400'	8,500'	Canyon Lakes

ACCESS: From Fort Collins 22 miles north on Highway 287, 24 miles west on County Road 74E, Red Feather Lakes Road, at Livermore; 3 miles south on County Road 162, Manhattan Road. **ATTRACTIONS:** The trail follows an old logging road through a lodgepole pine forest. About 1/2 mile from the trailhead, visitors can view the South Lone Pine drainage to the north. The south side of Molly Lake is on private property. Please do not trespass. **USE:** Low. **ACTIVITIES:** HIKING, HORSES. **USGS:** RED FEATHER LAKES QUAD. **MAP:** 3.
* Trail distance is 1.5 miles for Molly Lake, 2 miles for Elkhorn Creek and 3.5 miles for County Road #74 E.

953	Trail Name	Map Loc.	Distance	Difficulty	Beginning Elev.	Ending Elev.	Ranger District
Trail Number	**North Lone Pine**	F 2	3.7 mi	Easy	9,400'	10,700'	Canyon Lakes

ACCESS: 20 miles north from Fort Collins on Highway 287 and then 28 miles west on County Road 74E, Red Feather Lakes Road, and County Road 162, Deadman Road. Pine Creek to the Bald Mountain Jeep Trail near Middle Bald Mountain. Lodgepole pine gives way to spruce and fir as the elevation increases. Views of the Red Feather Lakes area are available at the upper end of the trail **ATTRACTIONS:** Scenery. **USE:** Low. **ACTIVITIES:** HIKING, MTN BIKING, HORSES. **USGS:** RED FEATHER LAKES QUAD. **MAP:** 3.

956	Trail Name	Map Loc.	Distance	Difficulty	Beginning Elev.	Ending Elev.	Ranger District
Trail Number	**Killpecker**	F 2	4.0 mi	Easy	9,300'	10,700'	Canyon Lakes

ACCESS: 21 miles north of Fort Collins on Highway 287 continue west on County Road 74E at Livermore for 23 miles and then west on County Road 162 for 8 miles. **ATTRACTIONS:** The trail ascends from the Deadman Road to the Bald Mountain Jeep Trail at Middle Bald Mountain. **USE:** Low. **ACTIVITIES:** HIKING, MTN BIKING, HORSES, MOTORCYCLES, ATV's. **USGS:** SOUTH BALD MOUNTAIN QUAD. **MAP:** 3.

CAMPGROUNDS LOCATED IN MAP 3

Map No.	Name	Fee	No. of Units	Max. Length	Elev.	Toilets	Water	Ranger District
1.	North Fork Poudre	$	9	30'	9,150'	Yes	No	Canyon Lakes
2.	Bellaire Lake	$	26	60'	8,650'	Yes	Yes	Canyon Lakes
3.	West Lake	$	29	50'	8,200'	Yes	Yes	Canyon Lakes
4.	Dowdy Lake	$	62	40'	8,140'	Yes	Yes	Canyon Lakes

MAP 4

No trail descriptions for this map

No campgrounds located on this map

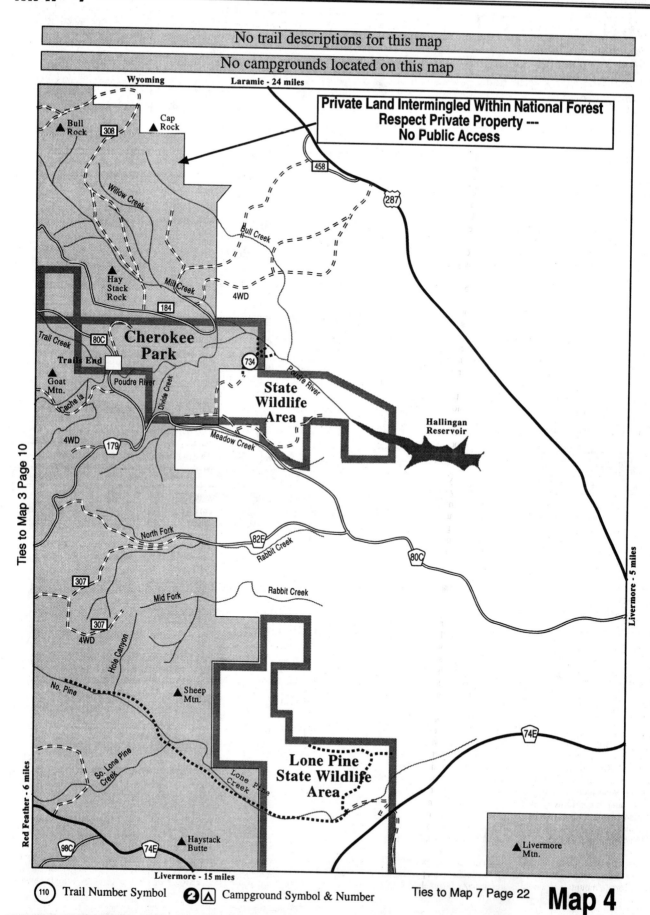

Private Land Intermingled Within National Forest Respect Private Property --- No Public Access

Wyoming

Laramie - 24 miles

▲ Bull Rock

308

▲ Cap Rock

Willow Creek

458

287

Bull Creek

▲ Hay Stack Rock

Mill Creek

184

4WD

Trail Creek

80C

Cherokee Park

Trails End

734

Poudre River

▲ Goat Mtn.

Cache la

Poudre River

Divide Creek

State Wildlife Area

Meadow Creek

Hallingan Reservoir

4WD

179

Ties to Map 3 Page 10

North Fork

82E

Rabbit Creek

80C

307

Mid Fork

Rabbit Creek

307

4WD

Hole Canyon

No. Pine

▲ Sheep Mtn.

74E

Red Feather - 6 miles

So. Lone Pine Creek

Lone Pine Creek

Lone Pine State Wildlife Area

98C

74E

▲ Haystack Butte

▲ Livermore Mtn.

Livermore - 5 miles

Livermore - 15 miles

(110) Trail Number Symbol   ❷ ⛺ Campground Symbol & Number   Ties to Map 7 Page 22   **Map 4**

Backpacking Is Freedom

Backpacking offers freedom to the forest traveler. You have no worries, other than your own. You become part of a scenic landscape and survive in a primitive environment with few modem conveniences. Self-sufficient, yes, but with this freedom goes an individual responsibility to care for the environment and respect the rights of those you meet along the way and those who follow you.

Backpacking is not limited to supermen and superwomen. However, it does require physical stamina and a genuine liking for the isolation in the remote country. Overnight backpacking trips should be undertaken only by those who have hiked easier mountain or forest trails and are familiar with back-packing techniques.

Leave No Trace

For thousands of years our wild lands have existed in a complex ecological interrelationship. This interrelationship can be easily upset or even destroyed. Once damaged, some plants and soils may not recover in our lifetime. Today, nature is struggling in many backcountry areas to cope with results of unacceptable backpacking, overnight camping techniques and heavy use.

Unappreciative or uninformed backpackers who have no enthusiasm for preserving the land are now in the minority. Even so, many backcountry areas are "camped out." Firewood is scarce or nonexistent. Unnatural fire blackened rocks and fire scars dot the landscape, and small green trees and ground cover are gone. In many areas, the streams are no longer safe for drinking. Several groups of people camping around the same lake lower the quality of the "backcountry experience" through noise and visual pollution.

Laws and regulations are being enforced to correct and eliminate these situations, but cooperation, proper attitudes, and voluntary actions of visitors are better ways to preserve the land. The concept of taking only pictures and leaving only footprints evolves from backpacker awareness.

Special regulations

Permits are required in some areas of the backcountry. Permits can be obtained from the local offices of the land managing agency. The permit must be obtained in advance and must be in your possession during your visit.

Group Size

In many backcountry areas the maximum number of people in a group is restricted. Large groups are destructive. Check to determine allowable group size.

Trail Courtesy

When hiking it's quite possible you may encounter trail riders along with pack stock. Livestock are easily spooked from unseen sources, it is best to make your presence known. When stock approaches, step off on the lower side of the trail while the stock passes. Be courteous in sharing the trail with others.

Fishing and Hunting

Write in advance to the State Game and Fish Department for fishing and hunting rules and licenses. Fishing and hunting are authorized under State regulations. Check with the local Game and Fish official before entering areas to fish or hunt because regulations vary. In every jurisdiction, the "plinking" gun used to destroy chipmunks, song birds, and other wildlife is held in contempt, and it is usually illegal.

Pets

Regulations differ on taking pets into the backcountry so check with the local Ranger regarding restrictions. Remember: dogs and cats are predators by nature and will instinctively chase forest birds and animals,horses and dogs don't mix. Physical restraint of the dog is necessary and bears and dogs don't mix.

You know your pet but other persons do not. Many areas have leash restrictions, especially on or within specified distances (usually 300 feet) of well traveled trails or in heavily used areas. Show respect for other persons and wildlife by keeping your pet under physical restraint or better yet, you might consider leaving your pet at home.

Awareness and Techniques

Backpacker awareness means understanding how you fit into the backcountry scene and not leaving evidence of your visit. If such awareness were practiced, all visitors would have the same opportunity to experience the natural scene. This awareness is intended to create backpacker recognition of the fragility of backcountry areas and a personal commitment to the care and wise use of this land.

If we could look back at the Rockies, the Southwest, or the Lake States in 1830, we would see a land devoid of cities, roads and vehicles, inhabited only by Indians and mountain men. When traveling the backcountry, the mountain man's priorities were adventure, monetary gain and personal survival. Today's visitors to the backcountry seek solitude, primitive recreation, and natural scenery. Yesterday's mountain man left no sign of his presence in Indian country. Today, backpackers should leave no signs of their presence so that the next person can enjoy a natural scene and the solitude it portrays. You must tread lightly so nature can endure and replenish.

Trip Planning

The first step of awareness and backpacking technique is planning your trip. As one of numerous visitors in the backcountry, plan your trip carefully to protect yourself as well as the environment.

For a carefully planned trip consider:

a. Maps to plan access, take-off and return points, route of travel, approximate camping areas and points of attraction to visit.
b. Proper lightweight equipment to safely cope with the elements and your recreational pursuits.
c. Food for the entire trip, packed in lightweight containers such as plastic bags.
d. Number of persons in the party and their abilities.
e. Regulations and restrictions that may be applicable.

Experience will help you refine planning skills, equipment, and techniques. However, evenings at home with how-to-do-it books, practice in putting up tents or shelters from ground cloths, and trying out dehydrated foods or home recipes will spark the imagination and eliminate some mistakes.

What You Need For:

Camping: Pack a tent or tarp for a shelter, sleeping bag, foam pad, lightweight stove, cooking utensils, dishes and cutlery and a small flashlight with extra batteries and bulb. Food should include snacks for the trail.

Clothing: Bring slacks or jeans - -2 pairs, long sleeved cotton shirts, at least 2 wool shirts or a sweater, parka windbreaker, wool socks - - 2 changes, underwear, camp shoes and socks, rain gear, rain shirt, poncho or nylon raincoat and handkerchiefs.

First Aid Kit: (you can make your own) Bring adhesive bandages, compresses, 4-inch elastic bandages, triangular bandage, antiseptic, aspirin, eye wash, adhesive tape, insect repellent, sunscreen lotion, mole-skin for blisters, tweezers and chapped lip medication.

Hiking: Wear footwear with eyelet's and lacing. Most backpackers prefer 6 to10-inch laced boots with rubber or synthetic soles. Footwear should be "broken in" and fit comfortably over two pairs of socks, one light and one heavy. Take extra laces. Pack a pair of soft sole shoes. After a day of hiking, they will feel comfortable as well as being less damaging to campsite vegetation.

Personal Sanitation: Carry a lightweight shovel or trowel and toilet paper.

Extra Comfort: Bring dark glasses, rope (nylon cord), knife, small pliers, waterproof matches, biodegradable soap, a towel, needle, thread and safety pins.

When to Travel

Time your trip according to climactic conditions. For example, in California's Hoover Wilderness backpacking season is about 2 months long — July and August. Even then, the hiker and camper should be prepared for all kinds of weather including rain, summer blizzards, extreme cold and heavy winds. In the Colorado mountains, conditions are usually favorable for traveling June 15 to October 1, but in the Northern Rockies, the best time for a trip is between

Continued on Page 21

MAP 5

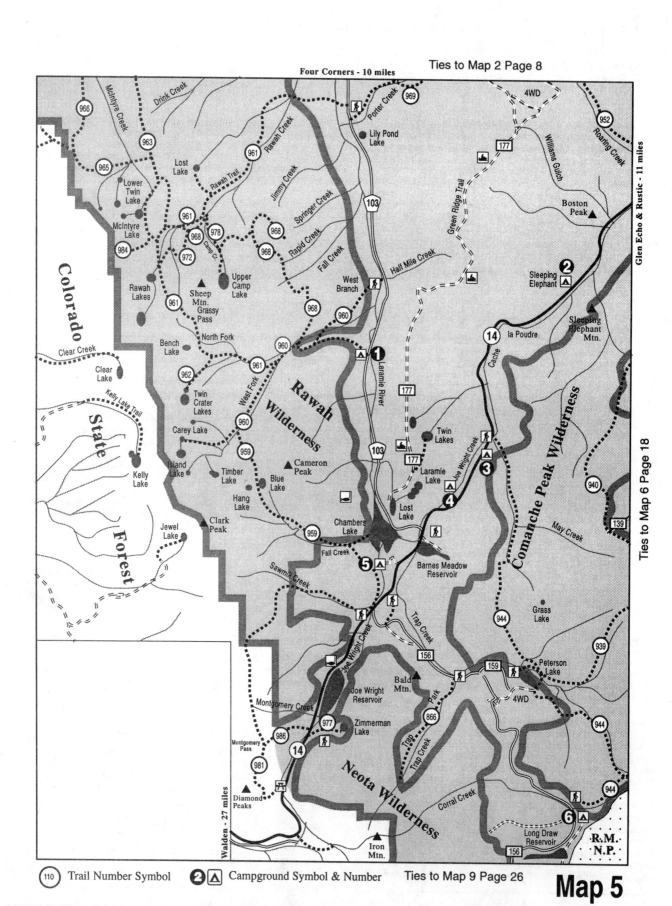

Four Corners - 10 miles

Glen Echo & Rustic - 11 miles

Colorado

State

Forest

Rawah Wilderness

Comanche Peak Wilderness

Ties to Map 6 Page 18

Neota Wilderness

Walden - 27 miles

R.M.
N.P.

⑪⓪ Trail Number Symbol	②△ Campground Symbol & Number	Ties to Map 9 Page 26

Map 5

MAP 5

866 Trail Number	Trail Name	Map Loc.	Distance	Difficulty	Beginning Elev.	Ending Elev.	Ranger District
	Trap Park	E 4	3.5 mi	Easy	9,975'	10,600'	Canyon Lakes

ACCESS: 9 miles north on Highway 287 from Fort Collins, 63 miles west on Highway 14, 4 miles south on County Road 156, Long Draw Road. **ATTRACTIONS:** The Trap Park trail follows an old road from the Long Draw Road up into Trap Park. The park is a beautiful valley bounded on the southwest by Iron Mountain and on the south by Flattop Mountain. The Neota Wilderness surrounds the park to the west, east, and south. Special regulations apply within the wilderness. From the end of the park, visitors can climb to the peak of Iron Mountain which offers panoramic views in all directions. **USE:** Low. **ACTIVITIES:** HIKING, HORSES. **USGS:** CHAMBERS LAKE QUAD. **MAP:** 5.

940 Trail Number	Trail Name	Map Loc.	Distance	Difficulty	Beginning Elev.	Ending Elev.	Ranger District
	Zimmerman	E 4	12.0 mi	Mod/Diff	10,400'	7,600'	Canyon Lakes

ACCESS This trail lies 4.5 miles east of Chambers Lake, running from the west tip of the Crown Point Road north to Highway 14. Travel west from Fort Collins about 30 miles up the Poudre Canyon to the Pingree Park Road, Forest Route 131. Follow this road south to the Crown Point Road. Turn west on Crown Point, and the trailhead is at the end of the road. The Poudre Canyon is nine miles to the north. The Flowers Trail is 3 miles to the south. Access can also be gained by following a 1.5 miles trail from the timber harvest roads to the Zimmerman Trail. **ATTRACTIONS:** Popular for horse back riding, trail offers fishing opportunities off Sheep, West Fork and East Fork Creeks. The last 3 miles of the trail are steep, so set your pace accordingly. **USE:** Light. **ACTIVITIES:** HIKING, HORSES, FISHING. **USGS:** COMANCHE PEAK, CHAMBERS LAKE, BOSTON PEAK, KINIKINIK QUADS. **MAPS:** 5 & 6.

944 Trail Number	Trail Name	Map Loc.	Distance	Difficulty	Beginning Elev.	Ending Elev.	Ranger District
	Big South	E 4	12.0 mi	Moderate	8,450'	9,400'	Canyon Lakes

ACCESS: Travel west on Highway 14 from Fort Collins about 50 miles. The trailhead is one mile past Poudre Falls on the south side of the road. The trailhead is visible from the road. Use the parking along the road **ATTRACTIONS:** Whether you are backpacking or just out for a hike, good fishing can be reached by this trail. Long Draw Reservoir can be reached from the south end of this trail, however route is not suitable for horses. **NARRATIVE:** This trail used to have a bridge over the Cache La Poudre River to allow easy access to the Peterson Lake area from the Flowers Trail. The bridge was destroyed and has not been replaced yet, making the only way across the river at this point a matter of some risk. At times, the river is low enough that people can wade or jump across exposed rocks to cross the river, but at other times, the water level is too high for safe crossing. Check the river flows before making a commitment to cross at that point. Designated camping -- Camp in numbered sites, stoves only. **USE:** Heavy. No Horses. **ACTIVITIES:** HIKING, MTN BIKING, FISHING. **USGS:** CHAMBERS LAKE, BOSTON PEAK QUADS. **MAP:** 5.

959 Trail Number	Trail Name	Map Loc.	Distance	Difficulty	Beginning Elev.	Ending Elev.	Ranger District
	Blue Lake	E 4	7.2 mi	Easy	9,520'	10,800'	Canyon Lakes

ACCESS: 9 miles north on Highway 287 from Fort Collins and then west on Highway 14 for 62 miles. **ATTRACTIONS:** The trail follows the Fall Creek drainage to Blue Lake. Panoramic views of the Mummy Range and Rocky Mountain National Park are available in the Blue Lake/Hang Lake area. Horses are prohibited on the Blue Lake Trail from May 15 to September 15 annually to help reduce adverse impacts on the trail and lake area. Overnight use is not permitted in the Blue Lake Closure Area. Overuse and abuse in the past have resulted in severe damage to the areas around the lakes and a closure has been implemented to allow the area to recover. Camping along the trail may be limited due to the high popularity of this drainage. **USE:** High. No horses. **ACTIVITIES:** HIKING, FISHING. **USGS:** CHAMBERS LAKE, CLARK PEAK, RAWAH LAKES QUADS. **MAP:** 5.

CAMPGROUNDS LOCATED IN MAP 3							
Map No.	Name	Fee	No. of Units	Max. Length	Elev.	Toilets	Water	Ranger District
1.	Tunnel	$	49	40'	8,600'	Yes	Yes	Canyon Lakes
2.	Sleeping Elephant	$	15	20'	7,850'	Yes	Yes	Canyon Lakes
3.	Big South	$	4	25'	8,440'	Yes	No	Canyon Lakes
4.	Aspen Glen	$	8	30'	8,660'	Yes	Yes	Canyon Lakes
5.	Chambers Lake	$	52	30'	9,200'	Yes	Yes	Canyon Lakes
6.	Long Draw	$	25	30'	10,030'	Yes	Yes	Canyon Lakes

MAP 5

960 Trail Number	Trail Name	Map Loc.	Distance	Difficulty	Beginning Elev.	Ending Elev.	Ranger District
	West Branch	E 3	6.98 mi	Moderate	8,560'	11,000'	Canyon Lakes

ACCESS: 9 miles north from Fort Collins on Highway 287, continue 59 miles west on Highway 14, and then 6 miles north on County Road 103, Laramie River Road. **ATTRACTIONS:** This trail climbs through lodgepole pine/aspen forest, following the west bank Laramie River. It crosses the river about 3 miles from the trailhead, then continues along the west branch of the river to Island and Carey Lakes. There is no bridge at the river crossing, but it is easily fordable during most of the summer and fall. Past its headwater crossing the West Branch Trail is joined by the Rawah Trail on its way down from Grassy Pass. It also intersects the Blue Lake Trail at Preacher's Camp, about a mile below the lakes. Limited camping is available along the first 3 miles of trail and at the lakes. Camp 200 feet from lakes, no fires in alpine area. **USE:** High. **ACTIVITIES:** HIKING, HORSES, FISHING. **USGS:** RAWAH LAKES, BOSTON PEAK QUADS. **MAP:** 5.

960/961 Trail Number	Trail Name	Map Loc.	Distance	Difficulty	Beginning Elev.	Ending Elev.	Ranger District
	Rawah South Access/Portal	E 3	9 mi	Moderate	8,560'	11,200'	Canyon Lakes

ACCESS: 9 miles north from Fort Collins on Highway 287, continue 59 miles west on Highway 14, and then 6 miles north on County Road 103, Laramie River Road. **ATTRACTIONS:** Elevation gain: 2,600 feet, loss 380 feet (Rawah Lake No. 1) High point: 11,250 feet. Allow 5 1/2 to 6 hours one way. Situated northwest of Rocky Mountain National Park in the heart of the Medicine Bow Range. The trail to West Branch Lakes goes southwest to a basin just above timberline and the trail to Camp Lakes traverses northward along the eastern boundary of the Wilderness. The climb over 11,200 foot Grassy Pass to the Rawah Lakes traverses especially attractive country. In addition to the pleasing vistas, many varieties of wild flowers grow along the route and large gray rabbits frequent the slopes near the pass. Many side trips are possible, including climbs to Twin Crater Lakes and Bench Lake. The main trail continues beyond the lowest Rawah Lake to Sandbar and Camp Lakes. **NARRATIVE:** Keep left and walk at a very gradual grade for two thirds mile to the footbridge over the North Fork of the Laramie River. Continue to the junction of the West Branch Trail. Turn right and climb at a moderate grade through deep woods. One mile from the junction cross two large streams. Although no foot bridges span the flows, the fords are not difficult. Switchback up, then hike at a gradual grade to the junction of the Twin Crater Lakes Trail at 5.0 miles. The route to these two lakes gains 650 feet of elevation in 1.2 miles. Keep right and soon begin dropping into an open grassy swale and at 5.8 miles ford a shallow stream. The faint route to Bench Lake heads southwest from this crossing. You can look down onto the lake from Grassy Pass if you want to study the terrain before making this side trip. Climb moderately through the meadows and clusters of trees toward the pass for one mile from the crossing then rise above timberline in two gradual switchbacks. Traverse up the slope to the pass. Veer slightly left toward a large rock marker then traverse cross-country along the grassy slope at a moderate uphill grade, bearing very slightly left. You are aiming for a point just to the southwest of the low summit ahead to the north. Pass a tarn on your left and come to a level area. Continue across it for a short distance to a viewpoint 350 feet above Rawah Lake No. 3. The cross-country route to the cirque that holds Rawah Lake No. 4 heads southwest from here. If you want to continue the hike to the lower Rawah Lakes, descend north from the overlook on a trail along the west side of the low ridge above the lake.
USE: Moderate. **ACTIVITIES:** HIKING, HORSES. **USGS:** RAWAH LAKES QUAD. **MAP:** 5.

961 Trail Number	Trail Name	Map Loc.	Distance	Difficulty	Beginning Elev.	Ending Elev.	Ranger District
	Rawah	D 3	13.7 mi	Moderate	8,400'	10,500'	Canyon Lakes

ACCESS: 9 miles north on Highway 287 from Fort Collins. Continue 59 miles west on Highway 14 up the Poudre Canyon and then 11 miles north on County Road 103, Laramie River Road. **ATTRACTIONS:** The trail follows Rawah Creek from the floor of the Laramie River Valley, past the Rawah Lakes, and over Grassy Pass to its junction with the West Branch Trail along the Laramie River. Visitors may directly access the Camp Lake Trail, McIntyre Lake Trail, North Fork Trail, and Link Trail from the Rawah Trail. In the lower elevations, the trail climbs through lodgepole pine, which gradually gives way to spruce and fir at the higher elevations. The section of the trail between Rawah Lake #2 and the south side of Grassy Pass is all above treeline. The trail leaves the trailhead and proceeds along an easement through private property. Please do not trespass. This trail is popular with horsemen and hikers in the summer months and is utilized by hunters in the fall. In late spring and early summer, sections of the trail may be boggy. **USE:** Moderate. **ACTIVITIES:** HIKING, HORSES. **USGS:** RAWAH LAKES QUAD. **MAP:** 5.

MAP 5

965 Trail Number	Trail Name	Map Loc.	Distance	Difficulty	Beginning Elev.	Ending Elev.	Ranger District
	Medicine Bow	D 3	11.5 mi	Moderate	10,500'	9,870'	Canyon Lakes

ACCESS: 9 miles north of Fort Collins on Highway 287. Continue 59 miles west on Highway 14 and then take County Road 103, Laramie River Road. Access the Medicine Bow Trail from the Rawah, Link, or McIntyre Creek/Ute Pass Trails. **ATTRACTIONS:** The trail follows the Medicine Bow Divide from Ute Pass south to join the Link Trail near McIntyre Lake. The trail is difficult to follow at some points. Views of North Park. the Laramie River Valley, Wyoming and Rocky Mountain National Park are available at numerous places along the trail. **USE:** Low. **ACTIVITIES:** HIKING. **USGS:** RAWAH LAKES, SHIPMAN MTN., JOHNNY MOORE MTN. QUADS. **MAPS:** 1 & 5.

977 Trail Number	Trail Name	Map Loc.	Distance	Difficulty	Beginning Elev.	Ending Elev.	Ranger District
	Zimmerman Lake	D 4	0.9 mi	Moderate	10,019'	10,480'	Canyon Lakes

ACCESS: 9 miles north from Fort Collins on Highway 287 and then 65 miles west on Highway 14. **ATTRACTIONS:** This trail follows an old road between Highway 14 and Zimmerman Lake. It climbs through a spruce/fir forest to the lake, which rests at the foot of the cliffs forming the northern border of the Neota Wilderness. Zimmerman Lake is one of the greenback cutthroat fisheries in this area. **USE:** Moderate. **ACTIVITIES:** HIKING, MTN BIKING, HORSES, FISHING. **USGS:** CHAMBERS LAKE, CLARK PEAK QUADS. **MAP:** 5.

986 Trail Number	Trail Name	Map Loc.	Distance	Difficulty	Beginning Elev.	Ending Elev.	Ranger District
	Montgomery Pass	D 4	1.8 mi	Moderate	10,040'	11,000'	Canyon Lakes

ACCESS: Drive 9 miles north on Highway 287 from Fort Collins. Then continue 65 miles west on Highway 14. The trail is located across Highway 14 from the Zimmerman Lake parking lot. **ATTRACTIONS:** The trail climbs from the Poudre Canyon to the top of Montgomery Pass. The pass offers magnificent views of North Park, sections of the Neota Wilderness to the south, and parts of the Rawah Wilderness to the north and west. Access to Diamond Peaks is available. **USE:** Low. **ACTIVITIES:** HIKING, MTN BIKING, HORSES. **USGS:** CLARK PEAK QUAD. **MAP:** 5.

Poudre River -- L. Carr USFS

MAP 6

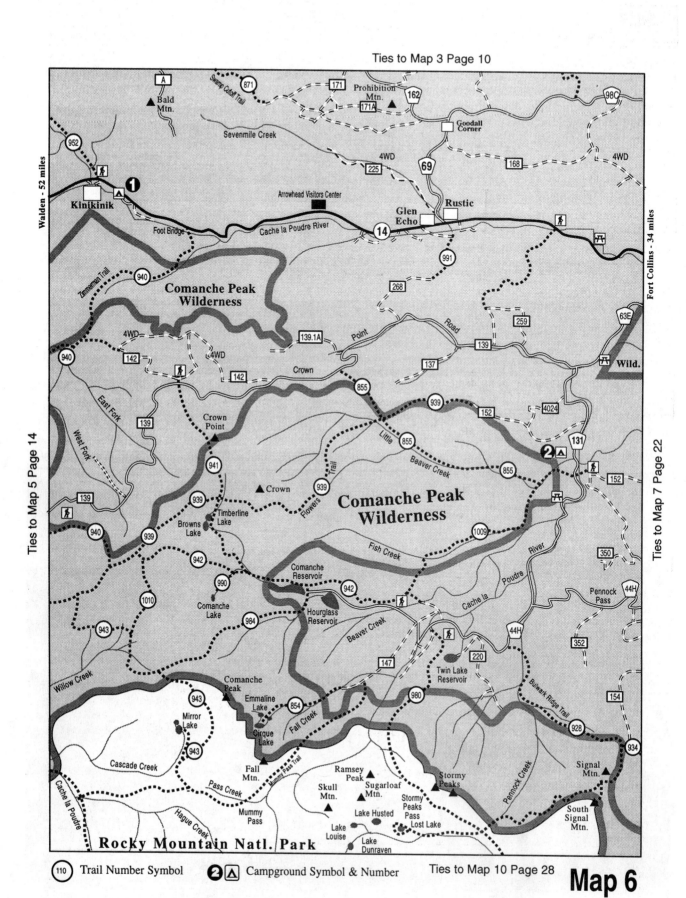

Ties to Map 3 Page 10

A

▲ Bald Mtn.

871   171   Prohibition Mtn. ▲   162   98C
171A

Swamp Cutoff Trail

Sevenmile Creek

Goodall Corner

952

4WD   69   168   4WD
225

Walden - 52 miles

Arrowhead Visitors Center   Glen Echo   Rustic

Kinikinik   14

Foot Bridge   Cache la Poudre River   991

Fort Collins - 34 miles

940   268

Comanche Peak Wilderness

4WD   139.1A   Point   Road   259   63E

940   142   4WD   Crown   139   137   Wild.

Zimmerman Trail   142   855

East Fork   855   939   152   4024

139   131

West Fork   Crown Point ▲   941   Little   855   855   152

Ties to Map 5 Page 14   939   ▲ Crown   Beaver Creek   2 ▲

Comanche Peak Wilderness   350

939   Timberline Lake   Fish Creek   1009   Poudre River

Browns Lake   942   Cache la   Pennock Pass   44H

940   990   984   Comanche Reservoir   942   220   352

1010   Comanche Lake   Hourglass Reservoir   Beaver Creek   147   44H

943   980   154

Willow Creek   Comanche Peak ▲   Emmaline Lake   854   Twin Lake Reservoir   Bulwark Ridge Trail   928

943   Mirror Lake   Cirque Lake   Fall Creek   Ramsey Peak ▲   Stormy Peaks   Signal Mtn. ▲

943   Cascade Creek   Fall Mtn. ▲   Skull Mtn. ▲   Sugarloaf Mtn. ▲   Stormy Peaks Pass   South Signal Mtn.

Cache la Poudre   Pass Creek   Mummy Pass Trail   Lake Husted   Lost Lake   934

Hague Creek   Mummy Pass   Lake Louise   Lake Dunraven   Pennock Creek

Rocky Mountain Natl. Park

Ties to Map 7 Page 22

110 Trail Number Symbol   2 ▲ Campground Symbol & Number   Ties to Map 10 Page 28   **Map 6**

MAP 6

854 Trail Number	Trail Name	Map Loc.	Distance	Difficulty	Beginning Elev.	Ending Elev.	Ranger District
	Mummy Pass-Emmaline Lake	G 4	6.0 mi	Moderate	8,800'	11,000'	Canyon Lakes

ACCESS: From Ft. Collins, take Highway 14 west for 35 miles and turn south on the Pingree Park Road. After 15 miles, turn right on the Tom Bennett Campground Road. Park at the campground or at the small trailhead 1/4 mile further.
ATTRACTIONS: The Emmaline Lake Trail is a road for the first two miles. Please stay on the road, do not wander into private land in Pingree Park. The trail leads 4 miles further to Emmaline and Cirque Lakes east of Comanche Peak. The Mummy Pass Trail heads 2 miles into Rocky Mountain National Park and leaves the Emmaline Lake Trail just past where it crosses Fall Creek. Interesting loops can be made with other Rocky Mountain National Park trails. **USE:** Moderate.
ACTIVITIES: HIKING, HORSES, FISHING. **USGS:** COMANCHE PEAK QUAD. **MAP:** 6.
DISTANCE: 6.0 MILES TO EMMALINE LK. 2.0 MILES TO MUMMY PASS

855/1009 Trail Number	Trail Name	Map Loc.	Distance	Difficulty	Beginning Elev.	Ending Elev.	Ranger District
	Little Beaver-Fish Creek	G 4	6.0 mi	Moderate	8,360'	9,000'	Canyon Lakes

ACCESS 1: Travel west from Ft. Collins about 35 miles on Highway 14. Take the Pingree Park Road south. After 6 miles, turn right on the Jack's Gulch Road. Park at trailhead at end of Jack Gulch Campground. Hike 2/3 mile to access trail to Little Beaver. The Fish Creek Trail can be reached from the northeast corner of Sky Ranch. **ACCESS 2:** On Pingree Park Road about 1.2 miles south of Flowers Road turn off very small parking area,1 or 2 cars. **ATTRACTIONS:** This trail is nice for hiking, horseback riding and fishing. Both trails may be brushy in places and require some bushwhacking. **USE:** Light. **ACTIVITIES:** HIKING, HORSES, FISHING. **USGS:** PINGREE PARK, RUSTIC QUADS. **MAP:** 6.
DISTANCE: LITTLE BEAVER: 6.0 MILES. FISH CREEK: 4.0 MILES.

939 Trail Number	Trail Name	Map Loc.	Distance	Difficulty	Beginning Elev.	Ending Elev.	Ranger District
	Flowers	G 3	11.5 mi	Mod/Diff	9,000'	9,300'	Canyon Lakes

ACCESS: This trail runs from the west tip of the Flowers four-wheel drive road and goes southwest to Peterson Lake. Take Highway 14 west out of Fort Collins to the townsite of Egger, approximately 30 miles. Turn south at the Pingree Park Road and go about 5 miles to the Old Flowers Road, also labeled the Jack's Gulch Road. This four-wheel drive road runs 3.6 miles to the Flowers Trail, 1.8 miles after Bedsprings Spring. **ATTRACTIONS:** Formerly used as a wagon trail in the late 1800's, this trail bisects the northern half of the District and ties into numerous other trails in the Crown Point -Mummy Range area. A canteen is recommended. **USE:** Moderate. **ACTIVITIES:** HIKING, HORSES. **USGS:** CHAMBERS LAKE, COMANCHE PEAK, KINIKINIK, RUSTIC QUADS. **MAP:** 5 & 6.

941 Trail Number	Trail Name	Map Loc.	Distance	Difficulty	Beginning Elev.	Ending Elev.	Ranger District
	Browns Lake	F 3	5.0 mi	Mod/Diff	9,500'	10,480'	Canyon Lakes

ACCESS: Go west out of Fort Collins on Highway 14 about 30 miles to Egger. Go south on Road 64E or the Pingree Park Road. Follow this road approximately 5 miles until you come to the Crown Point Road, which forks off to the west. Take the Crown Point Road another 12 miles or so, and the Brown's Lake Trailhead is off to the south side of the road. It is 4 miles to the lakes. Brown's Lake can also be reached from the Beaver Creek Trail and trailhead, 5 miles to the southeast.
ATTRACTIONS: This trail offers good fishing. You can walk on to Timberline Lake, Comanche Reservoir, and Hourglass Reservoir, which are within 2 miles of Brown's Lake. Please note that no motor vehicles are allowed on the trail due to resource damage. **USE:** Heavy. **ACTIVITIES:** HIKING, HORSES, FISHING. **USGS:** COMANCHE PEAK, KINIKINIK QUADS. **MAP:** 6.

Map No.	Name	Fee	No. of Units	Max. Length	Elev.	Toilets	Water	Ranger District
CAMPGROUNDS LOCATED IN MAP 6								
1.	Big Bend	$	9	20'	7,700'	Yes	No	Canyon Lakes
2.	Jacks Gulch	$	71	50'	8,000'	Yes	Yes	Canyon Lakes

MAP 6

942 Trail Number	Trail Name	Map Loc.	Distance	Difficulty	Beginning Elev.	Ending Elev.	Ranger District
	Beaver Creek	G 4	7.0 mi	Moderate	9,400'	11,120'	Canyon Lakes

ACCESS: Ride up the Poudre Canyon, Highway 14, 30 miles. Take Forest Route 131, Poudre Park Road, south off the highway for approximately 15 miles until you reach the Tom Bennett Campground. A rough road is located on the north side of the campground. Follow this road about 4 miles and you will reach the Beaver Creek Trailhead. **ATTRACTIONS:** This trail travels by Comanche Reservoir. So if you like to fish, go prepared. The trail offers a nice 1 mile jaunt west of Comanche Reservoir to Comanche Lake for the hiker and the equestrian as well. Continuing on across the Flowers Trail, you will connect with the Brown's Lake Trail. **USE:** Moderate. **ACTIVITIES:** HIKING, HORSES, FISHING. **USGS:** COMANCHE PEAK, PINGREE PARK QUADS. **MAP:** 6.

943 Trail Number	Trail Name	Map Loc.	Distance	Difficulty	Beginning Elev.	Ending Elev.	Ranger District
	Mirror Lake	F 4	9.0 mi	Moderate	11,320'	11,200'	Canyon Lakes

ACCESS: This trail lies due north of the northern border of Rocky Mountain National Park. **ACCESS 1:** From the Flowers Trail, the Mirror Lake Trail breaks off to the south and leads to Mirror Lake. **ACCESS 2:** Go west out of Fort Collins on Highway 14 about 30 miles to Egger. Take the Pingree Park Road south, Forest Route 131, and travel 15 miles to the Tom Bennett Campground. A rough road takes off north out of the campground. Two miles down this road is the Beaver Creek Trail. Follow this trail past Comanche Reservoir about 7 miles and the Mirror Lake Trail will be visible on the south side of the trail. **ATTRACTIONS:** This trail ventures into Rocky Mountain National Park before reaching Mirror Lake. Be sure to check with the Park for regulations. The Mummy Pass Trail is also accessed from the trail. There is fishing in this area, and the views are spectacular. **NARRATIVE:** See trail numbers #942 and #939. **USE:** Moderate. **ACTIVITIES:** HIKING, HORSES, FISHING. **USGS:** COMANCHE PEAK QUAD. **MAP:** 6.

952 Trail Number	Trail Name	Map Loc.	Distance	Difficulty	Beginning Elev.	Ending Elev.	Ranger District
	Roaring Creek	F 3	5.3 mi	Easy/Mod	7,700'	10,000'	Canyon Lakes

ACCESS: 9 miles north from Fort Collins on Highway 287 and then 48 miles west on Highway 14. **ATTRACTIONS:** The Roaring Creek Trail #952 runs from the Poudre Canyon to the Bald Mountain Jeep Trail #177. Portions of the first mile of this trail offer panoramic views of the Poudre Canyon. The first mile is steep, but the remainder of the trail is flat or gently rising. **USE:** Low. **ACTIVITIES:** HIKING, MTN BIKING, HORSES. **USGS:** DEADMAN, BOSTON PEAK, KINIKINIK QUADS. **MAPS:** 2, 5 & 6.

984 Trail Number	Trail Name	Map Loc.	Distance	Difficulty	Beginning Elev.	Ending Elev.	Ranger District
	Hourglass	F 4	4.0 mi	Difficult	8,400'	12,000'	Canyon Lakes

ACCESS: This trail lies directly north of Rocky Mountain National Park. The trail begins at Beaver Creek Trailhead and connects with Mirror Lake Trail. Go west from Fort Collins on Highway 14, 30 miles to Egger. Take Forest Route 131, Pingree Park Road, south for another 15 miles. Take a rough road west for 2 miles to the Beaver Creek trailhead. It is 2 miles west on the Beaver Creek Trail to the Hourglass Trail. **ATTRACTIONS:** This trail passes by Hourglass Reservoir and connects with the Mirror Lake Trail, which travels through Rocky Mountain National Park. Many hiking and fishing opportunities are available off this trail, all with striking scenery. **USE:** Moderate. **ACTIVITIES:** HIKING, HORSES, FISHING. **USGS:** DEADMAN, BOSTON PEAK, KINIKINIK QUADS. **MAPS:** 2, 5 & 6.

Continued From Page 13

July 15 and September 15. If you go into the high country too early, snow may interfere with travel, streams tend to be high and difficult to cross, fishing may be poor, and meadows and trails are apt to be soft and subject to damage. July and August are subject to intense afternoon thunder and lightning storms in the alpine areas. August and early September often provide the best weather for travel in the high country, with little bother from insects.

Locating a Campsite

Check with the local ranger for suitable camping areas; then plan your trip to avoid areas that need to recover from overuse.

a. If other parties are close to where you want to camp, move on or choose your campsite so that terrain features insure privacy. Trees, shrubs, or small hills will reduce noise substantially. Out of respect for nearby campers keep the noise level low at your campsite.

b. Use an existing campsite whenever possible, in order to reduce human impact. If selecting a new campsite, choose a site on sandy terrain or the forest floor rather than the lush but delicate plant life of meadows, stream banks, fragile alpine tundra and other areas that can be easily trampled or scarred by a campfire.

c. Camp at least 200 feet away from water sources, trails, and "beauty spots" to prevent water and visual pollution.

d. Take a little extra time to seek out a more secluded area. It will increase your privacy and that of other visitors.

e. Arrange the tents throughout the campsite to avoid concentrating activities in the cooking area.

f. Avoid trenching around your tent, cutting live branches or pulling up plants to make a park-like campsite. If you do end up clearing the area of twigs, or pine cones, scatter these items back over the campsite before you leave.

g. A backcountry campsite should be reasonably organized. If you have laundry to dry or equipment to air out, try to make sure these items are not in sight of other campers or hikers.

h. Leave the area as you found it, or in even better condition.

Travel Light

Experienced backpackers pride themselves on being able to travel light. Rugged, sure footed backpackers will seriously explain that they cut towels in half and saw the handles off toothbrushes to save ounces. They measure out just the right amount of food needed and put it in plastic bags, which are light. They carry scouring pads with built-in soap, to eliminate dish soap and a dishcloth.

How much should you carry? It all depends on your physical condition and experience, the terrain to be covered, the length of the trip, and the time of year. The average is 30 pounds for women (maximum 35) and 40 pounds for men (maximum 50). When figuring weight, count all items the cup on your belt, the camera around your neck and the keys in your pockets.

Backcountry Travel

Travel quietly in the backcountry, avoid clanging cups, yells, and screams. Noise pollution lessens the chance of seeing wildlife and is objectionable to others seeking solitude. However, in "grizzly country" noises may keep the bears away.

a. Wear "earth colors" to lessen your visual impact, especially if you are traveling in a group. However, during hunting season a blaze orange hat and vest are advisable for your personal safety.

b. When tracking wildlife for a photograph or a closer look, stay downwind, avoid sudden motions and never chase or charge any animal. Respect the needs of birds and animals for undisturbed territory. Some birds and small animals may be quite curious, but resist the temptation to feed them. Feeding wildlife can upset the natural balance of their food chain, your leftovers may carry bacteria harmful to them.

c. Stay on the designated path when hiking existing trails. Shortcutting a switchback or avoiding a muddy trail by walking in the grass causes unnecessary erosion and unsightly multiple paths. In the spring, travel across snow and rocks as much as possible; high mountain plants and soil are especially susceptible to damage during a thaw.

d. If you choose a route without trails, do not mark the trees, build rock piles, or leave messages in the dirt. A group should spread out.

e. Hike in groups of 4 to 6 people at most, 4 is the best number, especially during off-trail travel. In case of sickness or injury, one person can stay with the victim while two people go for help. Use your judgment in breaking your group into smaller units to reduce visual impact and to increase individual enjoyment and self-reliance.

f. Pick up any litter along the route; have one pocket of your pack available for trash.

g. Avoid removing items of interest (rocks, flowers, wood or antlers). Leave these in their natural state for others to see.

h. Allow horses plenty of room on trails. Horses may be frightened by backpack equipment. It is best to move off the trail. Everyone in your group should stand off to the downhill side of the trail. Avoid sudden movements as horses pass.

i. Help preserve America's cultural heritage by leaving archeological and historical remains undisturbed, encourage others to do the same and report your discoveries to the local ranger.

Campfires and Stoves

The mountaineer's decision to have a campfire was frequently influenced by the friendliness of the Indians. Today, your most important consideration should be the potential damage to the environment. A stove leaves no trace.

a. You should use a campfire infrequently and only when there is abundant dead wood available on the ground. Be very critical about the necessity for campfires. In many areas, wood is being used faster than it grows. In over camped areas or near timberline, choose an alternate campsite or use a portable stove.

b. In all areas fires should be completely out before you abandon the campsite. In some areas campfires are prohibited by regulations. Check with the public land management agency for local regulations.

If you do build a campfire remember:

All fires must be attended. Be aware of overuse. If your fire pit is full of wood ash or our cooking area unnecessarily trampled, move your campsite to lessen the camping scar.

a. Fires should be built away from tents, trees, branches and underground root systems.

b. Campfires should never be built on top of the forest floor. If there is a ground cover of needles and decomposed matter be sure to dig through it to the soil.

c. Be sure the fire pit is large enough to prevent the possibility of the fire spreading.

d. Do not build fires on windy days when sparks might be dangerous, especially when the countryside is dry.

Types of Fires

If you come upon a fire ring in the backcountry and the surrounding area has not been over camped, make use of it. However, fires should not be ringed with rocks as this permanently blackens them. When there is no existing fire ring, use one of the following three types of fires to assure little impact.

Flat Rock Method: Spread several inches of carefully gathered bare soil on top of a flat rock over an area slightly larger than the fire will occupy, then build your fire as usual. Burn all wood completely. After the fire is out, crush and scatter any coals. After the soil is removed and the rock rinsed, the area will be virtually unscarred.

Pit Method: Remove sod or topsoil in several large chunks from a rectangular area, about 12"x24" (sufficient for a party of two). When excavating the pit, place the topsoil or sod neatly in a pile nearby, and the pile of bare soil around the fire pit area to avoid drying out surrounding vegetation. If bare soil is

Continued on Page 37

MAP 7

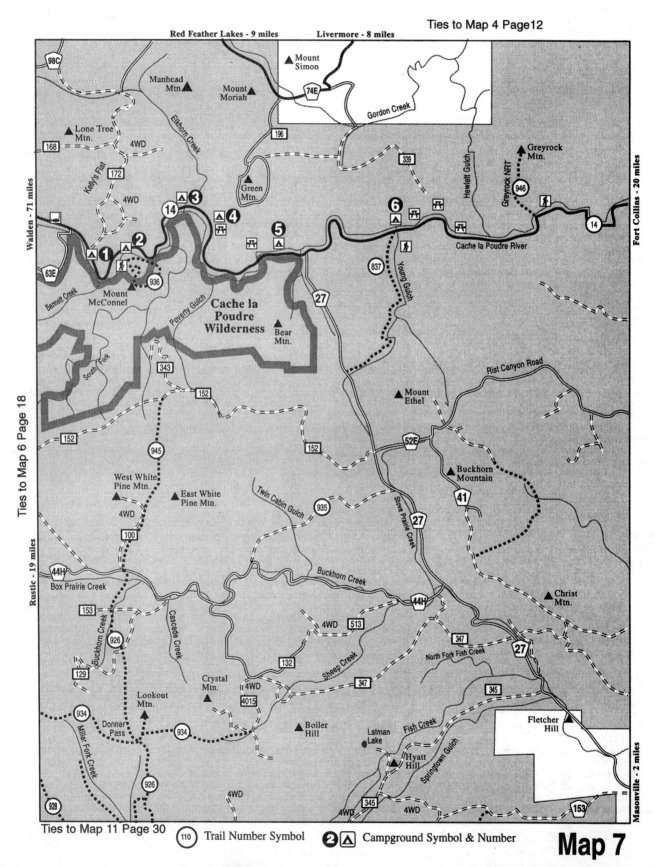

Ties to Map 4 Page12

Red Feather Lakes - 9 miles

Livermore - 8 miles

Walden - 71 miles

Fort Collins - 20 miles

Ties to Map 6 Page 18

Rustic - 19 miles

Masonville - 2 miles

Mount Simon

Manhead Mtn.

Mount Moriah

Lone Tree Mtn.

Green Mtn.

Greyrock Mtn.

Kelly's Flat

Mount McConnel

Bennett Creek

South Fork

Poverty Gulch

Cache la Poudre Wilderness

Bear Mtn.

Cache la Poudre River

Greyrock NRT

Hewlett Gulch

Young Gulch

Mount Ethel

Rist Canyon Road

Buckhorn Mountain

Gordon Creek

West White Pine Mtn.

East White Pine Mtn.

Twin Cabin Gulch

Stove Prairie Creek

Christ Mtn.

Box Prairie Creek

Buckhorn Creek

Buckhorn Creek

Cascade Creek

Crystal Mtn.

Sheep Creek

North Fork Fish Creek

Lookout Mtn.

Donner Pass

Boiler Hill

Latman Lake

Fish Creek

Fletcher Hill

Miller Fork Creek

Hyatt Hill

Springtown Gulch

98C, 168, 172, 4WD, 4WD, 196, 339, 63E, 936, 837, 27, 343, 152, 152, 152, 52E, 945, 4WD, 100, 44H, 153, 926, 129, 934, 934, 926, 928, 4WD, 513, 132, 4WD, 4015, 347, 347, 345, 345, 4WD, 4WD, 153, 41, 27, 44H, 347, 27, 946, 14, 74E

⑩ Trail Number Symbol

❷ △ Campground Symbol & Number

Map 7

MAP 7

837 Trail Number	Trail Name	Map Loc.	Distance	Difficulty	Beginning Elev.	Ending Elev.	Ranger District
	Young Gulch	J 3	5.3 mi	Easy/Mod	5,900'	7,400'	Canyon Lakes

ACCESS: From Fort Collins go west 12 miles along Highway 14 to the Ansel Watrous campground. The dirt road opposite the campground on south side of highway, leads to Young's Gulch Trailhead. **ATTRACTIONS:** This trail is heavily used by mountain bikers. There are over 20 stream crossings. Not recommended for horses due to the heavy use by bicycles. The stream crossings do not have foot bridges so expect wet feet by the end of your hike. No camping within 1/4 mile of trail. **USE:** Heavy. **ACTIVITIES:** HIKING, MTN BIKING. **USGS:** POUDRE PARK QUAD. **MAP:** 7.

936 Trail Number	Trail Name	Map Loc.	Distance	Difficulty	Beginning Elev.	Ending Elev.	Ranger District
	Mt. McConnel Nat. Rec.	H 3	2.5 mi	Moderate	6,720'	7,520'	Canyon Lakes

ACCESS: Travel from Fort Collins west up Poudre Canyon on Highway 14 approximately 26 miles. Turn south over a bridge to day-use area at Mtn. Park. Park at day-use area and Mt. McConnel trailhead. **ATTRACTIONS:** This signed interpretive nature trail offers an enjoyable hike for the young and old. There is no drinking water available on the trail. The Nature Trail is part of the National Recreation Trail System. The Mummy Range is viewed from the summit of Mt. McConnel. The descending loops are steep and rocky. Take a canteen of water. The road is heavily used. **USE:** Heavy. **ACTIVITIES:** HIKING. **USGS:** BIG NARROWS QUAD. **MAP:** 7. TRAIL DISTANCES: KREUTZER -- 2.0 MILES. MT. MC CONNEL -- 2.5 MILES.

946 Trail Number	Trail Name	Map Loc.	Distance	Difficulty	Beginning Elev.	Ending Elev.	Ranger District
	Greyrock National Rec.	J 3	2. 8 mi	Moderate	5,560'	7,600'	Canyon Lakes

ACCESS: From Fort Collins, go 9 miles north on Highway 287 and then 17 miles west on Highway 14. **ATTRACTIONS:** The trail climbs from the Poudre Canyon to the base of Greyrock. From the top of Greyrock a view of the Poudre Canyon, the plains to the east, and the mountains to the west and south is available. The Greyrock Meadow Trail branches off of the main trail about 0.6 mile from the trailhead, climbs into Greyrock Meadow, and then joins the main trail again near the base of Greyrock. This is a good day hike. Hikers should carry their own drinking water. **USE:** High. **ACTIVITIES:** HIKING, MTN BIKING, HORSES. **USGS:** POUDRE PARK QUAD. **MAP:** 7. TRAIL DISTANCE: SUMMIT TRAIL --2.85 MILES ONE-WAY. GREYROCK MEADOW 4.48 MILES ONE-WAY.

CAMPGROUNDS LOCATED IN MAP 7 -- Page 22.								
Map No.	Name	Fee	No. of Units	Max. Length	Elev.	Toilets	Water	Ranger District
1.	Kelly Flats	$	23	40'	6,750'	Yes	Yes	Canyon Lakes
2.	Mountain Park	$	55	50'	6,650'	Yes	Yes	Canyon Lakes
3.	Dutch George	$	21	33'	6,500'	Yes	Yes	Canyon Lakes
4.	Narrows	$	9	30'	6,500'	Yes	No	Canyon Lakes
5.	Stove Prairie Landing	$	9	30'	6,000'	Yes	Yes	Canyon Lakes
6.	Ansel Watrous	$	19	30'	5,800'	Yes	Yes	Canyon Lakes

MAP 8

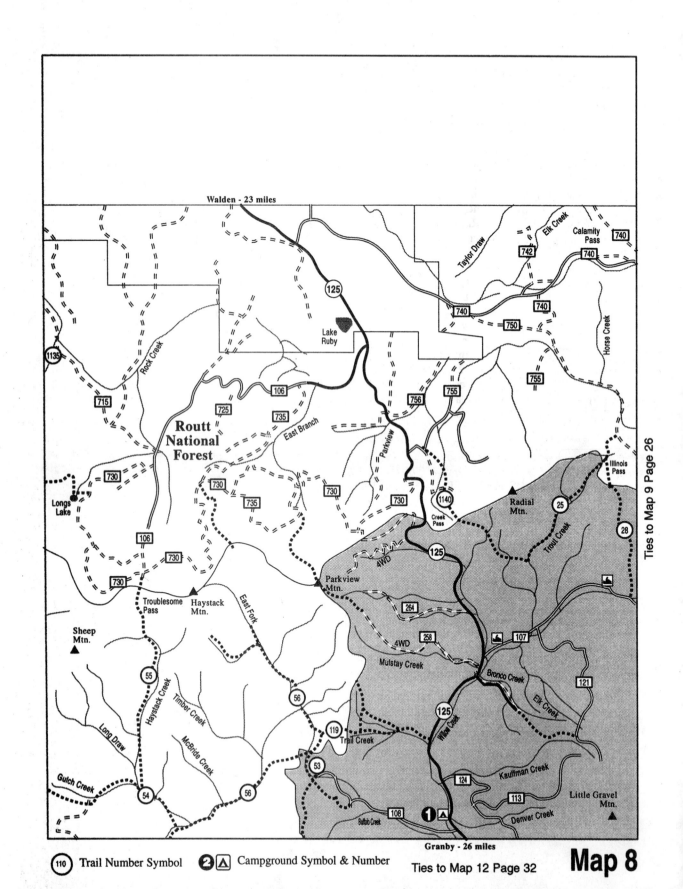

Walden - 23 miles

Routt
National
Forest

Lake
Ruby

1135
715
725
106
735
East Branch
730
730
735
730
730
Longs
Lake
106
730
730
Troublesome
Pass
Haystack
Mtn.
Sheep
Mtn.
East Fork
55
Haystack Creek
Timber Creek
McBride Creek
Long Draw
Gulch Creek
54
56
56
119
Trail Creek
53
108
Buffalo Creek

Parkview
Parkview
Mtn.
756
755
755
1140
Creek
Pass
4WD
125
264
4WD
258
Mulstay Creek
Radial
Mtn.
25
28
Illinois
Pass
Trout Creek
107
Bronco Creek
Elk Creek
121
125
Willow Creek
124
Kauffman Creek
113
Denver Creek
Little Gravel
Mtn.

Taylor Draw
Elk Creek
742
Calamity
Pass
740
740
740
740
750
Horse Creek

Ties to Map 9 Page 26

Granby - 26 miles

⑩ Trail Number Symbol   ❷△ Campground Symbol & Number

Ties to Map 12 Page 32

Map 8

MAP 8

Trail No.	Trail Name	Map Loc.	Distance	Difficulty	Beginning Elev.	Ending Elev.	Ranger District
---	**Parkview Mountain**	C 6	5.2 mi	Moderate	8,800'	12,296'	Sulphur

ACCESS: Proceed on Colorado 125 five miles south of Willow Creek Pass or 16 miles north of the junction with U. S. 40 to a sign stating Parkview Mountain Trail at a rough road on the west side of the highway. This spot is 0.3 mile south of a road heading east and downhill to the Vagabond Ranch, Stillwater Pass and the Lost Lake Trail. A few parking spaces are available on the highway shoulder near the trailhead or a short distance down the Stillwater Pass Road. **ATTRACTIONS:** From the summit of Parkview Mountain you can enjoy one of the most far-ranging panoramas in northern Colorado. Although the name of the massive peak probably applies to its location high above the southern end of North Park, you also will have views into Rocky Mountain National Park close to the east with an especially good sighting of the portion of Longs Peak between the Keyhole and the summit. The Park Range on the horizon to the northwest includes the Mt. Zirkel Wilderness and the Medicine Bow Range with the Rawah Wilderness in its center is on the skyline to the northeast. Below this latter chain is the Never Summer Range, the western boundary of Rocky Mountain National Park. The view extends south to the Holy Cross region and southwest to the area around Aspen and the distinctive level crest of the Flat Tops. **NARRATIVE:** The first 3.8 miles of the hike climbs in woods along a rough jeep road to timberline then the route begins winding cross-country up slopes of grass and wild flowers to the broad, gentle summit ridge. Carry drinking water as sources along the hike may not be dependable. Climb along the road for a short distance then curve left and walk above an old cabin. At 0.2 mile begin traveling to the west beside the edge of a small valley filled with brush and beaver ponds. Near its western end veer away from the swale and resume climbing in woods. The grade generally is moderate but periodically the road climbs at a somewhat steeper angle. At 1.8 miles come to a fork and turn right as indicated by the sign pointing to Parkview Mountain. Red metal tags imbedded in trees identify the correct route at several junctions further along the road. Pass the remains of a pole pen and at 2.3 miles keep left at a fork. Two-tenths mile further stay right and after another 0.2 mile keep left. The grade soon becomes very moderate as the route traverses to the north. Walk on the level for about one-quarter mile then watch for red tabs that mark a side road going left. This junction is across from a post on the right side of the main road. Turn left and follow this side road through more scenic woods. Walk beside a small creek then just beyond a large clearing come to timberline at 3.8 miles. Continue on the road, rising moderately along a slope brightened in midsummer with the rich yellow color of buttercups. Look back and note landmarks so on the return you will be able to locate the spot where the road reenters the woods. Begin climbing cross country toward the crest and as you gain elevation the waxy buttercup blossoms are replaced by tiny alpine forget-me-nots, the variety of Indian paintbrush with leaves of a yellow-green hue, gentians and other pert wild flowers. Seven-tenths mile from timberline come to the crest and hike along the ridge top at a very moderate grade. Do not walk along the snow just below the crest on the east side as it may be corniced. Soon you will be able to see the lookout house ahead to the northwest. Follow the path that traverses the western slope just below the crest for 0.2 mile to the summit. **USE:** Moderate. **ACTIVITIES:** HIKING, MTN BIKING, HORSES. **USGS:** PARKVIEW MTN., RADIAL MOUNTAIN QUADS. **MAP:** 8.

CAMPGROUNDS LOCATED IN MAP 8								
Map No.	Name	Fee	No. of Units	Max. Length	Elev.	Toilets	Water	Ranger District
1.	Denver Creek	$	25	25'	8,600'	Yes	Yes	Sulphur

MAP 9

See RMNP Map 3 (Page 62) for Rocky Mountain National Park Trails

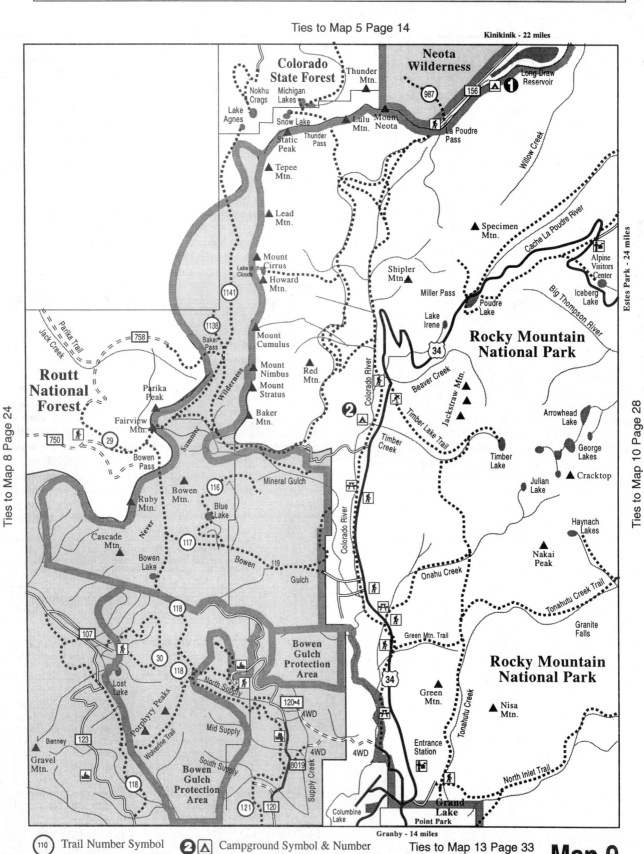

Ties to Map 5 Page 14

Kinikinik - 22 miles

Colorado State Forest

Neota Wilderness

Routt National Forest

Rocky Mountain National Park

Rocky Mountain National Park

Ties to Map 8 Page 24

Ties to Map 10 Page 28

Granby - 14 miles

110 Trail Number Symbol   2 △ Campground Symbol & Number   Ties to Map 13 Page 33

Map 9

MAP 9

Map No.	Name	Fee	No. of Units	Max. Length	Elev.	Toilets	Water	Ranger District
CAMPGROUNDS LOCATED IN MAP 9								
1.	Grand View	$	8	Tent	10,220'	Yes	Yes	Canyon Lakes
Rocky Mountain National Park								
2.	Timber Creek	$	100	30'	8,900'	Yes	Yes	RMNP

Map Symbol Explanation

🅿 Parking Area		Forest Service Facility	285 U.S. Highway		National Forest Area		
🅰 Picnic Area		Fishing Area	126 State Highway		Water		
Trail Head		Snowmobile Trail	258 County Highway		Trail		
Downhill Ski Area		Ice Skating Area	1606 Trail Number		River or Stream		
Boat Launch		Cross Country Ski Area	1105 Forest Service Road		Primary Road - Paved		
Bicycle Trail		Hunting	▲ Mountain		Improved Road - Unpav		
4WD Road		❶⬛ Campground	🅜 Colorado Trail		Unimproved Road - 4W		
Motorcycle Trail		Towns & Locales	Continental Divide Trail		Forest / Wilderness Boundary		

Middle Boulder Creek -- USFS

MAP 10

See RMNP Map 4 (Page 68) for Rocky Mountain National Park Trails

Ties to Map 6 Page 18

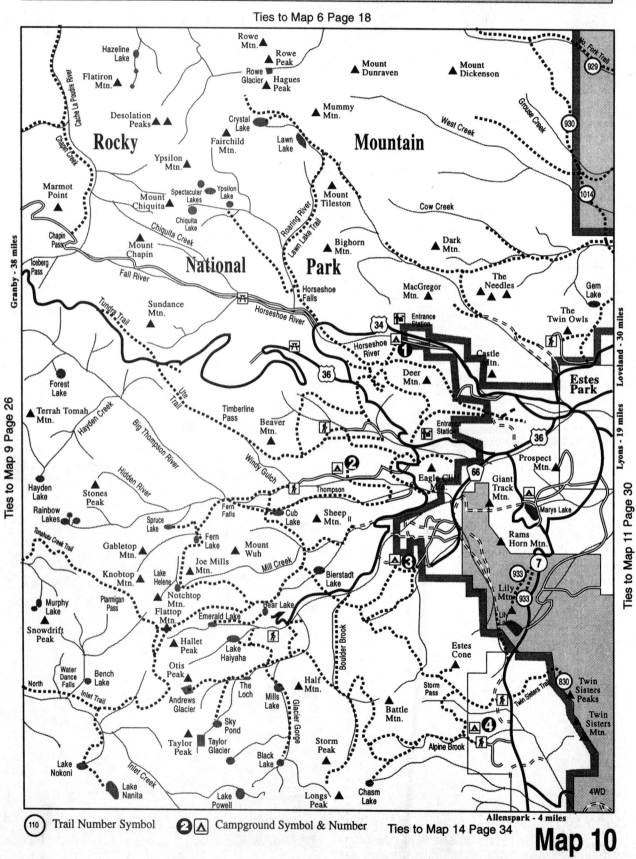

Granby - 38 miles

Ties to Map 9 Page 26

Loveland - 30 miles

Lyons - 19 miles

Ties to Map 11 Page 30

Allenspark - 4 miles

(110) Trail Number Symbol ❷△ Campground Symbol & Number Ties to Map 14 Page 34

Map 10

MAP 10

Trail Number 933	Trail Name	Map Loc.	Distance	Difficulty	Beginning Elev.	Ending Elev.	Ranger District
	Lily Mountain	G 6	1.7 mi	Easy/Mod	8,800'	9,786'	Canyon Lakes

ACCESS: From Estes Park, go south on Highway 7 about 5.8 miles. Just before the 6 mile marker there is a small pull-off area by the trailhead sign. Be careful because parking space is limited and the pull-off sits just off busy Highway 7.
ATTRACTIONS: Lily Mountain, at an elevation of 9,786 feet, is perhaps one of the easiest mountains to climb in the Estes Park area. With an elevation gain of 986 feet over 1.5 miles, the summit may be reached in 1 to 2 hours, depending on physical condition, number and length of stops, etc.. This is an ideal half-day excursion. There are numerous overlooks along the trail which provide magnificent views of the Estes valley. No water is available on the trail. **NARRATIVE:** The Lily Mountain trail first takes the hiker north on a gentle up and down trail that parallels Highway 7 well below the trail. After three-quarters of a mile, there is an unmarked trail junction. Take the left fork, the right fork leads to private land. Soon the trail cuts back south and switchbacks, sometimes steeply, towards the summit of Lily Mountain. The entire trail is well marked. The final one-eighth mile stretch to the summit is marked by cairns, small rock piles, and provides some fun rock scrambling. Instead of using the cairn trail, you can stay on the lower trail which ends in a saddle, a low point between mountain tops. The view from the summit of Lily Mountain makes the climbing effort worthwhile. Mountains are every-where! You behold a panoramic view including the Mummy Range, Continental Divide, Longs Peak, and Estes Cone, which are all part of Rocky Mountain National Park. **USE:** Moderate. **ACTIVITIES,** HIKING, HORSES. **USGS:** LONGS PEAK QUAD. **MAP:** 10.
NOTE: In the future the trail will be rerouted at Lily Lake.

CAMPGROUNDS LOCATED IN MAP 10								
Map No.	Name	Fee	No. of Units	Max. Length	Elev.	Toilets	Water	Ranger District
Rocky Mountain National Park								
1.	Aspenglen	$	54	30'	8,230'	Yes	Yes	RMNP
2.	Moraine Park	$	247	35'	8,000'	Yes	Yes	RMNP
3.	Glacier Basin -Group	$	---	---	8,600'	Yes	Yes	RMNP
4.	Longs Peak	$	26	30'	9,400'	Yes	Yes	RMNP

View From Loveland Pass -- USFS

MAP 11

No campgrounds located on this map

Ties to Map 7 Page 22

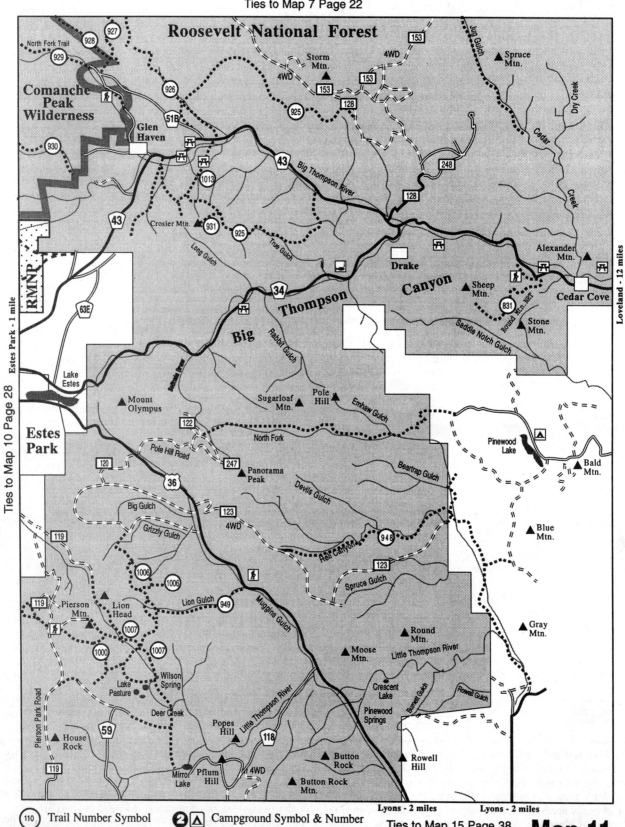

Roosevelt National Forest

Comanche Peak Wilderness

Glen Haven

Big Thompson River

Drake

Big Thompson Canyon

Cedar Cove

Alexander Mtn.

Spruce Mtn.

Storm Mtn.

Sheep Mtn.

Stone Mtn.

Estes Park - 1 mile

Ties to Map 10 Page 28

RMNP

Estes Park

Lake Estes

Mount Olympus

Sugarloaf Mtn.

Pole Hill

Pinewood Lake

Bald Mtn.

Panorama Peak

Blue Mtn.

Big Gulch

Grizzly Gulch

Pierson Mtn.

Lion Head

Gray Mtn.

Round Mtn.

Moose Mtn.

Wilson Spring

Lake Pasture

Deer Creek

House Rock

Crescent Lake

Pinewood Springs

Mirror Lake

Pflum Hill

Popes Hill

Button Rock

Button Rock Mtn.

Rowell Hill

Loveland - 12 miles

Lyons - 2 miles   Lyons - 2 miles

(110) Trail Number Symbol   ❷🏕 Campground Symbol & Number   Ties to Map 15 Page 38

Map 11

MAP 11

831 Trail Number	Trail Name	Map Loc.	Distance	Difficulty	Beginning Elev.	Ending Elev.	Ranger District
	Foothills Nature-Round Mtn.	J 5	**	Easy	5,743'	8,450'	Canyon Lakes

DISTANCE: 1.0 MILE TO NATURE TRAIL. 2.5 MILES TO SHEEP MTN. DIFFICULTY: EASY FOR NATURE TRAIL, THEN MODERATE. ACCESS 1: Access is approximately 12 miles west of Loveland on Highway 34. Just 3 miles up the Big Thompson Canyon you will see the trailhead on the south side of the road. Look for the Loveland Mountain Park area and the trailhead. **ACCESS 2: From Estes Park: Take Highway 34 east 15.5 miles to Viestenz-Smith Mountain Park. **ATTRACTIONS:** Horseback riding and viewing wildlife are among the activities to be enjoyed in the area. An interpretive program is available on the trail, discussing the ecology and environment. **NARRATIVE:** The trail begins with a self-guided nature trail run by the city of Loveland and then continues on Forest Service to the top of Sheep Mountain. Brochures are available in a box at the trailhead. There is some water available on the trail, but it is not treated for drinking. The nature trail section is a day use area. Camping and campfires are prohibited until you are on Forest Service land. Because of the many south facing slopes in this area, the trail is clear of snow and can be used year-round without much difficulty. The trail, which is maintained in cooperation with the City of Loveland, is a National Recreation trail. **USE:** Moderate. **ACTIVITIES:** HIKING, HORSES. **USGS:** DRAKE QUAD. **MAP:** 11.

929 Trail Number	Trail Name	Map Loc.	Distance	Difficulty	Beginning Elev.	Ending Elev.	Ranger District
	North Fork	H 5	4.5 mi	Easy	7,680'	8,800'	Canyon Lakes

ACCESS: From Loveland travel west on Highway 34 to Drake. Take the Devils Gulch road towards Glen Haven for 6 miles. At that point there will be a forest access sign directing you onto the Dunraven Glade Road. At the end of this road there will be a large parking area with a horse corral. From Estes Park travel east on the Devils Gulch road about 9 miles to the Forest Access Parking Area 2.25 miles sign. Turn north onto the Dunraven Glade road and at the end of this road there will be a large parking area with a horse corral. **ATTRACTIONS:** Once over and down the ridge west of the parking lot, the trail becomes fairly level as it parallels and crisscrosses the North Fork of the Big Thompson River. The trail remains close to the river until Deserted Village, 3 miles from the trailhead. Water is readily available for horses. Bridges are available for each river crossing. Please use these bridges to prevent the further erosion of the river banks at these crossings. Deserted Village is a large open meadow just after the sixth river crossing. Once an active hunting camp with several buildings, now there remains only one log structure. **NARRATIVE:** The trail runs through the Comanche Peak Wilderness area, and as with all Wilderness areas, mechanized equipment is prohibited to preserve the natural essence and character of the land. This means that mountain bikes and power tools are not allowed. Please minimize your impacts to the plants and trail and clear away all traces of your trip. The trail also runs across some sections of private land. Please respect the landowners and leave the area in the same or better shape as you pass through. The trail beyond Deserted Village leaves the side of the river for less than a mile before nearing the river again. The park's boundary is 1.5 miles beyond Deserted Village. When planning a trip into Rocky Mountain National Park, be sure to first check with the Park Headquarters for their regulations. **Connecting Trails:** North Fork (#929). Bulwark Ridge/ Signal Mountain Trail (#928). Indian Trail (#927). Donner Pass (#926) **USE:** Moderate. No Mechanized Equipment. **ACTIVITIES:** HIKING, HORSES. **USGS:** GLEN HAVEN, ESTES PARK QUADS. **MAPS:** 10 & 11.
Lost Lake, in Rocky Mountain National Park, is 4.25 miles from the boundary of the Park.

931 Trail Number	Trail Name	Map Loc.	Distance	Difficulty	Beginning Elev.	Ending Elev.	Ranger District
	Croiser Mountain	H 5	4.0 mi	Moderate	7,200'	9,250'	Canyon Lakes

ACCESS 1: From Estes Park take the Devil's Gulch road, County Road 43, to Glen Haven. From Loveland take Highway 34 west to Glen Haven via Drake and the Devil's Gulch Road. **ACCESS 2:** From Estes Park or Drake travel the Devil's Gulch Road. There is a large gravel pit on the south side of the road. A trail sign "Crosier Mtn." sits just up the hill from a gate. Please keep the gate closed. This trail is steep, but parking is ample. **ACCESS 3:** From Drake follow the Devil's Gulch road 2.2 miles west. Look for a small metal gate on the south side of the road. There is a small parking area near the gate. **ATTRACTIONS:** Numerous beautiful overlooks, especially from on top of Crosier Mountain, giant aspen stands and meadows filled with wildflowers highlight this series of trails. **NARRATIVE:** The trail from Glen Haven is the easiest to the summit. Starting at elevation 7,200 feet the summit is 2,050 vertical feet and 4 miles from the trailhead. Keep in mind the rule of thumb for estimating hiking time: plan 1 hour for every 1,000 vertical feet to be gained and add an hour for every 2-3 miles covered. The longest approach to the summit is Access 3 with 5 miles and 2,850 vertical feet to gain. **USE:** Moderate. **ACTIVITIES:** HIKING, MTN BIKING, HORSES. **USGS:** GLEN HAVEN QUAD. **MAP:** 11.

949 Trail Number	Trail Name	Map Loc.	Distance	Difficulty	Beginning Elev.	Ending Elev.	Ranger District
	Lion Gulch	H 6	2.75 mi	Moderate	7,360'	8,400'	Canyon Lakes

ACCESS 1: The Lion Gulch Trail is located on Highway 36, 13 miles west of Lyons and 7 miles east of Estes Park. From Lyons, travel west 13 miles, and you'll see the Lion Gulch trailhead sign on the south side of the road. **ACCESS 2:** From Estes Park, travel approximately 7 miles east on Highway 36. You will approach the Lion Gulch trailhead situated on the south side of the road. Park alongside the road near the trailhead. Be careful when entering traffic as visibility is limited. **ATTRACTIONS:** This trail is easily accessible off the highway. It follows Lion Gulch into the Homestead Meadows area. Interpretive signs at Homestead Meadows depicting the history of the pioneers of the area. **USE:** Moderate. **ACTIVITIES:** HIKING, MTN BIKING, HORSES. **USGS:** PANORAMA PEAK QUAD. **MAP:** 11.

MAP 12

CAMPGROUNDS LOCATED IN MAP 12

Map No.	Name	Fee	No. of Units	Max. Length	Elev.	Toilets	Water	Ranger District
1.	Sawmill Gulch	$	5	32'	8,780'	Yes	Yes	Sulphur

No trail descriptions for maps 12 and 13

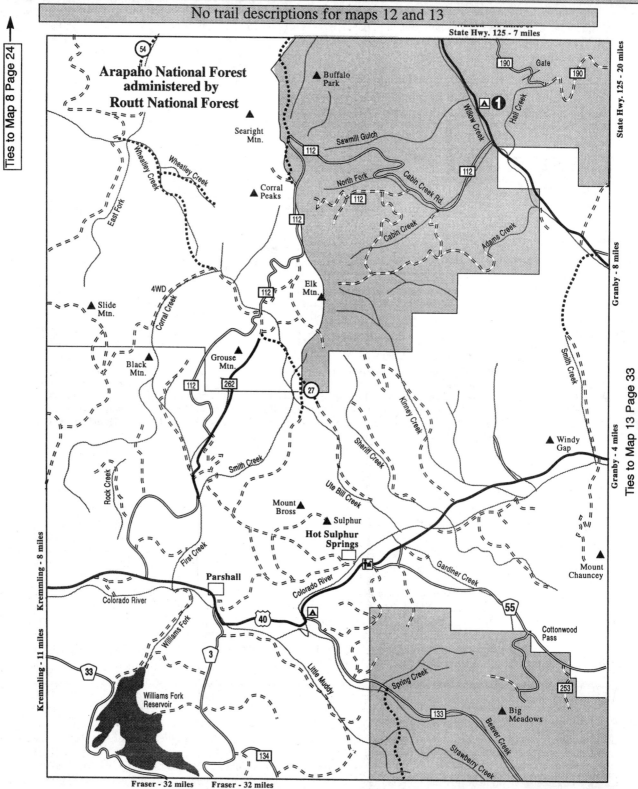

Ties to Map 8 Page 24

Arapaho National Forest administered by Routt National Forest

State Hwy. 125 - 7 miles

State Hwy. 125 - 20 miles

Granby - 8 miles

Ties to Map 13 Page 33

Granby - 4 miles

Buffalo Park

Searight Mtn.

Corral Peaks

Elk Mtn.

Slide Mtn.

Black Mtn.

Grouse Mtn.

Windy Gap

Mount Bross

Sulphur

Hot Sulphur Springs

Mount Chauncey

Parshall

Gardiner Creek

Cottonwood Pass

Big Meadows

Williams Fork Reservoir

Kremmling - 8 miles

Kremmling - 11 miles

Fraser - 32 miles

Fraser - 32 miles

Ties to Map 17 Page 40

(110) Trail Number Symbol   ❷ 🔺 Campground Symbol & Number

Map 12

MAP 13

Map No.	Name	Fee	# Units	Max. Length	Elev.	Toilets	Water	Ranger District
1.	Green Ridge	$	78	35'	8,400'	Yes	Yes	Sulphur
2.	Cutthroat Bay - Group	$	---	---	8,400'	Yes	Yes	Sulphur
3.	Stillwater	$	127	40'	8,350'	Yes	Yes	Sulphur
4.	Arapaho Bay	$	84	35'	8,320'	Yes	Yes	Sulphur
5.	Willow Creek	$	35	32'	8,130'	Yes	Yes	Sulphur

CAMPGROUNDS LOCATED IN MAP 13

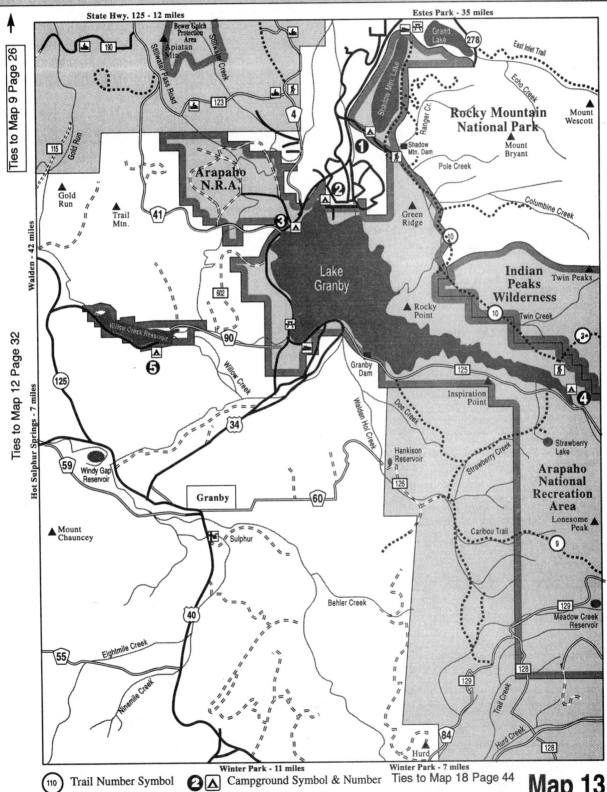

(110) Trail Number Symbol   ❷△ Campground Symbol & Number

Map 13

33

MAP 14

Campground information on page 36

Ties to Map 10 Page 28

Estes Park - 13 miles

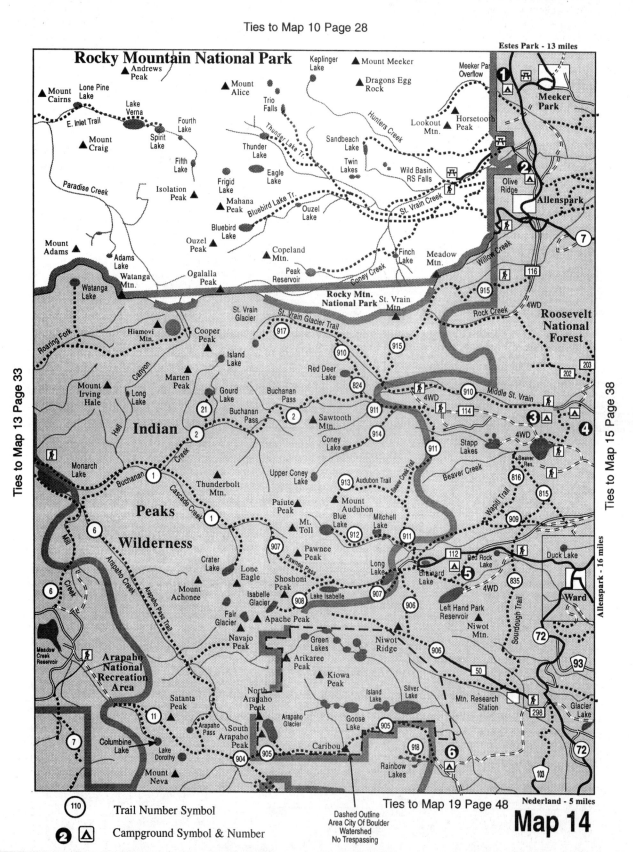

Ties to Map 13 Page 33

Ties to Map 15 Page 38

Allenspark - 16 miles

Ties to Map 19 Page 48

Nederland - 5 miles

(110) Trail Number Symbol

❷ ⬜ Campground Symbol & Number

Dashed Outline
Area City Of Boulder
Watershed
No Trespassing

Map 14

MAP 14

824/910/914 Trail Number	Trail Name	Map Loc.	Distance	Difficulty	Beginning Elev.	Ending Elev.	Ranger District
	Red Deer Cutoff	G 7	*	Moderate	9,800'	10,372'	Boulder

Buchanan Pass and Coney Lake Trails.

ACCESS: One mile west of Colorado 72 on FDR 114 to Camp Dick Campground. **ATTRACTIONS:** This portion of the wilderness is heavily used by horseback parties, fishermen, groups, and "loop" backpackers. All the trails have snowbanks across them in July. Please use caution when crossing them. Campsites at Red Deer and Coney Lakes are very limited to the surrounding topography. The upper switchbacks on Buchanan should be negotiated carefully if using horses. The last 0.5 mile to Coney Lakes involve bushwhacking, while the first 0.5 mile to Red Deer are steep and rocky. Those looking for solitude should camp away from the lakes and trails. If crossing Buchanan Pass, try to do so before midafternoon to avoid being above timberline when storms are most likely to boil up. Please observe the posted regulations and keep your traces of passing through to a minimum. Buchanan Pass mileages and elevations given, from the wilderness boundary. If you are planning to stay overnight camping permits are required.

NARRATIVE: This trail enters the wilderness. **USE:** Heavy to Extremely Heavy. **ACTIVITIES:** HIKING, HORSES, FISHING. **USGS:** ALLENSPARK, ISOLATION PEAK, MONARCH LAKE, WARD QUADS. **MAP:** 14.
* TRAIL DISTANCE:
RED DEER CUT-OFF: 1.0 MILE
BUCHANAN PASS: 3.5 MILES

906/907 Trail Number	Trail Name	Map Loc.	Distance	Difficulty	Beginning Elev.	Ending Elev.	Ranger District
	Niwot Ridge/Jean Luning	G 8	*	Easy	10,400'	11,670'	Boulder

ACCESS: County Road 102, Brainard Lake Road, off Colorado 72 at Ward. **ATTRACTIONS:** Wilderness regulations are posted at the trailheads. Please adhere to them. Camping and fires are prohibited on the Jean Luning Trail and in the high alpine forest/ tundra on Niwot Ridge in order to minimize user impacts. Since Niwot Ridge is a study area of extreme fragility and uniqueness, please remain on the trail and old road to the south. Information regarding the biosphere can be obtained by contacting the Arctic and Alpine Research Station (303-492-8841) or by going to the research station off the Rainbow Lakes Road. **NARRATIVE:** The Brainard area is the most heavily visited wilderness entrance in the Indian Peaks, with literally thousands of visitors in the area on any given summer weekend. Niwot Ridge Trail: This trail branches off of the Jean Luning Trail. Niwot Ridge has been designated an international biosphere by the U.N. for arctic and alpine research. Some of the strongest winds recorded in North America occur here, 160 miles per hour. **USE:** Extremely Heavy. **ACTIVITIES:** HIKING. **USGS:** WARD QUAD. **MAP:** 14.
* TRAIL DISTANCE:
NIWOT RIDGE: 2.5 MILES
JEAN LUNING: 1.2 MILES

907/908 Trail Number	Trail Name	Map Loc.	Distance	Difficulty	Beginning Elev.	Ending Elev.	Ranger District
	Pawnee Pass/Isabelle Glacier	G 8	*	Mod/Diff	10,400'	12,541'	Boulder

ACCESS: County Road 102, Brainard Lake Road, off Colorado 72 at Ward. **ATTRACTIONS:** While beautiful, these areas do not provide the solitude and tranquillity to be found elsewhere. When in the area, please remember to apply a wilderness ethic of being a visitor, leaving no trace of your passing across the land. Remember too that as you pass above timberline, altitude and weather become increasingly important factors. Do not assume that warm, sunny mornings at Brainard Lake will await you at higher elevations later in the day. Snowfields and loose rock can prove to be extremely hazardous to people unfamiliar with mountaineering techniques and equipment. Please use common sense. **NOTE:** The Ward quadrangle does not show the alignment of the Pawnee Pass Trail. As you near Lake Isabelle, DO NOT cross the creek but head east, right, into the woods at the trail junction. Above timberline, especially if it appears that thunderstorms are brewing. It is a good rule of thumb to leave the summit by 1:00 - 1:30, as thunderstorms begin to form in midafternoon. Camping above timberline is discouraged due to lack of shelter and the fragility of the tundra. Please stay on trails whenever possible. **NOTE:** The Ward quadrangle labels the Beaver Creek Trail as Buchanan Pass. **NARRATIVE:** The use of these trails is extremely heavy, carrying hundreds of people a day on weekends. The shore of Lake Isabelle and surrounding meadows have been noticeably scarred by campers, while the area between the lake and Pawnee Pass has been interlaced with "illegal" trails cut by over anxious hikers looking for shortcuts. Altitude is an increasing influence on the upper 2 miles of Pawnee Pass and final mile to Isabelle Glacier. Pawnee Pass is a popular "loop" trail for backpackers crossing the Divide to Monarch Lake. **USE:** Extremely Heavy. **ACTIVITIES:** HIKING. **USGS:** WARD, MONARCH LAKE QUADS. **MAP:** 14.
* TRAIL DISTANCE:
PAWNEE PASS: 4.5 MILES.
ISABELLE GLACIER: 2.0 MILES.

MAP 14

911/912/913 Trail Number	Trail Name	Map Loc.	Distance	Difficulty	Beginning Elev.	Ending Elev.	Ranger District
	****	G 8	****	Moderate	10,500'	10,400'	Boulder

ACCESS: County Road 102, Brainard Lake Road, off Colorado 72 at Ward. **ATTRACTIONS:** No camping, regulations are posted at the trailheads for your information. The Beaver Creek Trail is a popular horseback and backpacker "loop" trip to the Middle St. Vrain and Buchanan Pass. If using the Audubon Trail, carry water and foul weather gear. Be alert to weather changes anywhere. **NARRATIVE:** These trails are heavily traveled by a range of users from day hikers to horseback parties. At the Mitchell Lake parking lot, the Audubon/Beaver Creek trailhead is to the north, your right as you come in, while Mitchell is next to the bulletin board. The Mitchell trail ends at Blue Lake in the fragile forest/ tundra transition zone, while the Beaver Creek Trail climbs to 11,300 ft. before dropping down into the Middle St. Vrain drainage. A fork in this trail to the northwest leads to the summit of Mt. Audubon. **USE:** Extremely Heavy. **ACTIVITIES:** HIKING. **USGS:** WARD, ALLENSPARK QUADS. **MAP:** 14.

****TRAIL DISTANCE:

911	BEAVER CREEK:	6.0 MILES
912	MITCHELL LAKE:	2.5 MILES
913	MT. AUDUBON:	2.0 MILES (No Horses)<None>

915/917 Trail Number	Trail Name	Map Loc.	Distance	Difficulty	Beginning Elev.	Ending Elev.	Ranger District
	St Vrain Mtn & Glacier	G 7	*	Mod/Diff	9,800'	11,300'	Boulder

ACCESS: (St. Vrain Mountain) Take County Road 107 south out of Allenspark approximately 1 mile to FDR 116. Park here and begin the trail. **ATTRACTIONS:** This area is less heavily camped than most other regions in the wilderness. Little water is available on the St. Vrain Mountain Trail, and persons hiking this trail should be in good shape. It crosses through Rocky Mountain National Park near St. Vrain Mountain. Persons wishing to camp here should make sure they have a permit. The glacial remnants and moraines at the end or the St. Vrain Glaciers Trail are treacherous to hike/ski on. Please use common sense and use this region with a wilderness ethic in mind. Read and comply with the posted regulations. **NARRATIVE:** After 2.7 miles the St. Vrain Glacier Trail crosses and becomes faint up into a chaotic glacial moraine. Two lakes sit in cirques above the valley floor. St. Vrain Mountain Trail is very steep at its western end for 1.5 miles and at mile 2 at the eastern stretch. **USE:** Heavy to Extremely Heavy. **ACTIVITIES:** HIKING, HORSES. **USGS:** ALLENSPARK, ISOLATION PEAK QUADS. **MAP:** 14.

*TRAIL DISTANCES:

915	ST. VRAIN MOUNTAIN	6.5 MILES
917	ST. VRAIN GLACIERS	4.0 MILES

918 Trail Number	Trail Name	Map Loc.	Distance	Difficulty	Beginning Elev.	Ending Elev.	Ranger District
	Rainbow Lakes	G 8	1.0 mi	Easy	10,000'	10,300'	Boulder

ACCESS: County Road 116 west of Colorado 72 between Nederland and Ward. **ATTRACTIONS:** Principally a fishing area, Rainbow Lakes offers primitive camping facilities. The eastern end of the Arapaho Glacier Trail begins here. To preserve and enhance the wilderness character of the area, certain rules and regulations are posted. Please adhere to them. **NARRATIVE:** The trail head begins at the Rainbow Lakes Campground. It runs for 1 mile to the 8 small lakes. The area has been heavily used and campsites around the lakes show severe signs of deterioration. **USE:** Extremely Heavy. **ACTIVITIES:** HIKING, HORSES, FISHING. **USGS:** WARD QUAD. **MAP:** 14.

Map No. No	Name	Fee	No. of Units	Max. Lenght	Elev.	Toilets	Water	Ranger District
1.	Meeker Park Overflow	$	27	30'	8,600'	Yes	Yes	Boulder
2.	Olive Ridge	$	56	30'	8,350'	Yes	Yes	Boulder
3.	Camp Dick	$	41	55'	8,650'	Yes	Yes	Boulder
4.	Peaceful Valley	$	17	55'	8,500'	Yes	Yes	Boulder
5.	Pawnee	$	55	45'	10,350'	Yes	Yes	Boulder
6.	Rainbow Lakes	$	18	20'	10,000'	Yes	No	Boulder

CAMPGROUNDS LOCATED IN MAP 14

Continued From Page 21

not placed on top of sod surrounding the fire pit, then the sod should be kept moist. On breaking camp, both the bottom and sides of the fire pit should be cold to the touch. Remaining coals should be crushed to powder or paste before carefully replacing the dirt and sod. Make sure there are no soft spots in the fill-in fire pit that will sink with age. Also be sure to mold the edges of well-defined chunks of sod to assure a flat surface and to give the appearance that the earth has not been disturbed. Landscape the entire cooking area by scattering leaves, twigs or whatever originally covered the ground. It is worth the effort.

Surface Method: When there is abundant bare soil available without excavation (gopher holes, old stream beds, etc.) there should be no need to disturb the topsoil by digging a fire pit. Simply spread several inches of bare soil on the ground and build a fire as usual. As with the "flat rock method," all wood should be burned completely to ashes. Crush remaining coals and scatter ashes and bare soil once they have cooled. Be careful of scorching the topsoil, and landscape the cooking area before leaving.

Firewood Selection
a. Select your firewood from small diameter loose wood lying on the ground in order to insure complete, efficient burning.
b. Avoid breaking off branches, alive or dead, from standing trees. An area with discolored broken stubs and pruned trees loses much of its natural appearance.
c. Leave saws and axes at home because they leave unnatural and unnecessary scars and add weight to your pack.
d. The mark of an experienced backpacker is to use a stove when wood is not readily available or when an area could be easily damaged.
e. Firewood is often scarce near heavily used campsites so it should not be wasted on excessively large fires.
f. Scatter unused firewood- before leaving your campsite to preserve a natural appearance.

Extinguishing Fires
When preparing to leave the campsite use water and bare soil to douse the flames thoroughly. Feel the coals with your bare hands to be sure the fire is out, scatter and bury the ashes.

Human Waste
The proper disposal of human waste is important. For the benefit of those who follow, you must leave no evidence that you were there, and you must not contaminate the waters. Fortunately, nature has provided a system of very efficient biological "disposers" to decompose fallen leaves, branches, dead animals and animal droppings in the top 6 to 8 inches of soil. If every hiker cooperates, there will be no backcountry sanitation problems. The individual "cat method," used by most experienced backpackers is recommended.

The "cat method" includes the following steps:
a. Carry a light digging tool, such as an aluminum garden trowel. Select a screened spot at least 200 feet from the nearest water. Dig a hole 6 to 8 inches across. Try to remove the sod (if any) in one piece. Fill the hole with the loose soil, after use and then tramp in the sod. Nature will do the rest in a few days.
b. When hiking on the trail, burial of human waste should be well away and out of sight of the trail, with proper considerations for drainage. The cat method is unnecessary for urination; however, urinate well away from trails and water sources. Use areas that are well-hidden, but try to avoid vegetation because the acidity of urine can affect plant growth.
c. If you are traveling as a group, consider a toilet pit to minimize impact.
d. Burning of toilet paper is preferable to burying it since it does not decompose quickly. This is essential to prevent sanitation problems from heavy visitation. If you are up to the mountain man's style, use snow, leaves, and other natural substitutes in preference to toilet paper.
e. Tampons must be burned in an extremely hot fire to completely decompose. When not in grizzly bear country, they can be bagged and packed out. Never bury tampons because animals will dig them up.

Disposal of Camping Wastes
a. Tin cans, bottles, aluminum foil, and other "unburnables" should not be taken to the backcountry because they must be packed out.
b. Avoid the problem of leftover food by carefully planning meals. When you do have leftovers, carry them in plastic bags or burn them completely.
c. Waste water (dishwater or excess cooking water) should be poured in a corner of the fire pit to prevent attracting flies. If you cook on a stove, disperse water waste faraway from any body of water. Non-soluble food particles (macaroni or noodles) in dishwater should be treated like bulk leftovers. They should be either packed up and carried out or burned. Nothing should be left behind. Food scraps like egg and peanut shells and orange peels take a long time to decompose, and are eyesores to other hikers.
d. Fish intestines should be burned completely in a campfire. However, if there scavenger animals and birds around and not many remains, and if the area is lightly used, then the intestines can be scattered in discreet places to decompose naturally.
e. Use good judgment.

Drinking Water
Better to be safe than sorry. No matter how "pure" it may look, water from streams or lakes should be considered unsafe to drink until properly treated.

The most common disease associated with drinking water is giardiasis which is caused by ingesting the microscopic cyst form of the parasite giardia lambia. Flu-like symptoms appear 5 to 14 days after exposure and may last 6 weeks or more if untreated. Other disease causing organisms may also be present in untreated surface waters. Start each trip with a days supply of water from home or other domestic source. To replenish that supply, search out the best and cleanest source, then strain the water through a clean cloth to remove any suspended particles or foreign material. The best treatment then is to bring the water to a boil for 3 to 5 minutes. Cool overnight for the next day's supply.

Another solution is to treat with iodine water purification tablets or use an EPA approved filtered water purifier device.

Bathing and Washing
Although the mountain men weren't famous for their cleanliness, today's visitors like to bathe and wash their clothes. Be aware, however, that all soap pollutes lakes and streams. If you completely soap bathe, jump into the water first, then lather on the shore well away from the water, and rinse the soap off with water carried in jugs or pots. This allows the biodegradable soap to break down and filter through soil before reaching any body of water.
a. Clothes can be adequately cleaned by thorough rinsing. Soap is not necessary.
b. Too much soap in one place makes it difficult for soil to break it down. Therefore, dispose of soapy water in several places.
c. Do not use soap or dispose of soapy water in tundra areas; the soil layer is too thin to act as an effective filter and destruction of plant life usually results.

Safety and Emergency Precautions
For safety reasons, travel with a companion. Leave word at home and at your jumping-off place if a backcountry visitor register is provided. When you travel in a party, see to it that no one leaves the group without advising where they are going and for how long.

Watch out for loose or slippery rocks and logs, cliffs, steep grades, and inclined hard packed snow fields where a misstep can cause an uncontrolled slide or fall. Use your best judgment and never take chances.

In Case of Injury
Injury in remote areas can be the beginning of a real emergency. Stop immediately! Treat the injury if you can and make the victim comfortable. Send or signal for help. If you must go for help, leave one person with the injured. If rescue is delayed, make an emergency shelter. Don't move until help arrives

Continued on Page 47

MAP 15

No campgrounds located on this map

No trail descriptions for this map

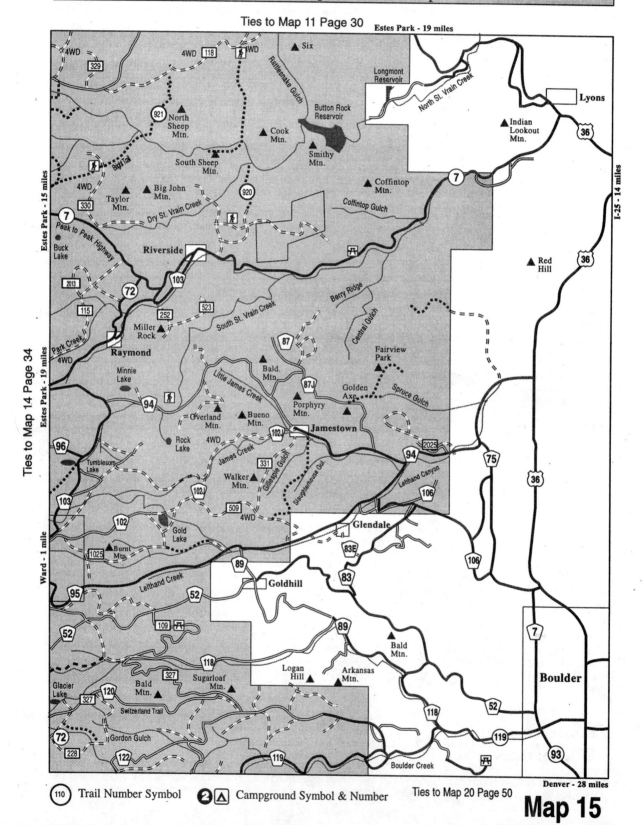

Ties to Map 11 Page 30

Estes Park - 19 miles

Trail Number Symbol — 110

Campground Symbol & Number — 2 △

Ties to Map 20 Page 50

Denver - 28 miles

Map 15

MAP 16

No campgrounds located in Arapaho/RooseveltNational Forest on this map

No trail descriptions for this map

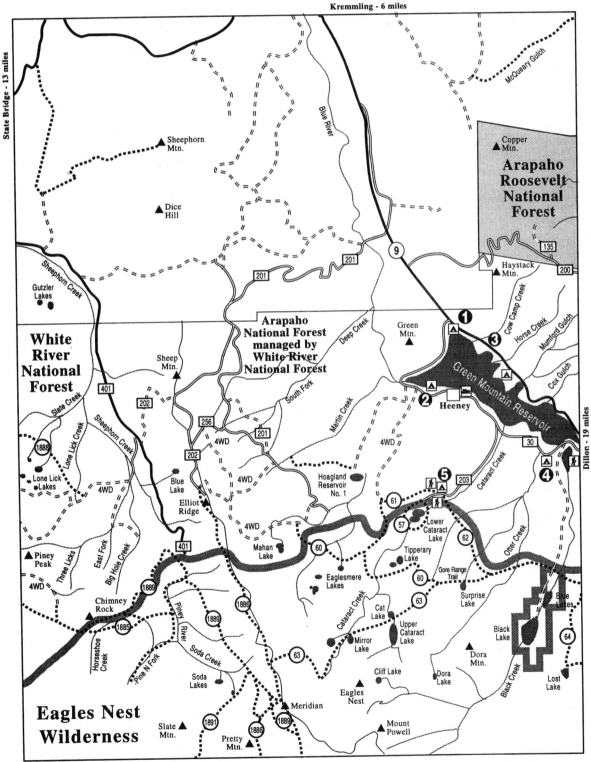

Kremmling - 6 miles

State Bridge - 13 miles

Ties to Map 17 Page 40

Dillon - 19 miles

Arapaho Roosevelt National Forest

Copper Mtn.

Sheephorn Mtn.

Dice Hill

Blue River

Sheephorn Creek

Gutzler Lakes

Haystack Mtn.

White River National Forest

Arapaho National Forest managed by White River National Forest

Sheep Mtn.

Green Mtn.

Heeney

Green Mountain Reservoir

Cow Camp Creek

Horse Creek

Mumford Gulch

Cox Gulch

Deep Creek

South Fork

Martin Creek

Slate Creek

Lone Lick Creek

Sheephorn Creek

Lone Lick Lakes

Blue Lake

Elliot Ridge

Hoagland Reservoir No. 1

Cataract Creek

Piney Peak

Mahan Lake

Eaglesmere Lakes

Lower Cataract Lake

Tipperary Lake

Gore Range Trail

Surprise Lake

Otter Creek

Blue Lakes

Chimney Rock

Piney River

Soda Creek

Cat Lake

Upper Cataract Lake

Black Lake

Dora Mtn.

Horseshoe Creek

Pine N Fork

East Fork

Big Hole Creek

Three Licks

Soda Lakes

Mirror Lake

Cliff Lake

Dora Lake

Black Creek

Lost Lake

Eagles Nest Wilderness

Slate Mtn.

Pretty Mtn.

Meridian

Eagles Nest

Mount Powell

(110) Trail Number Symbol   ❷△ Campground Symbol & Number

Map 16

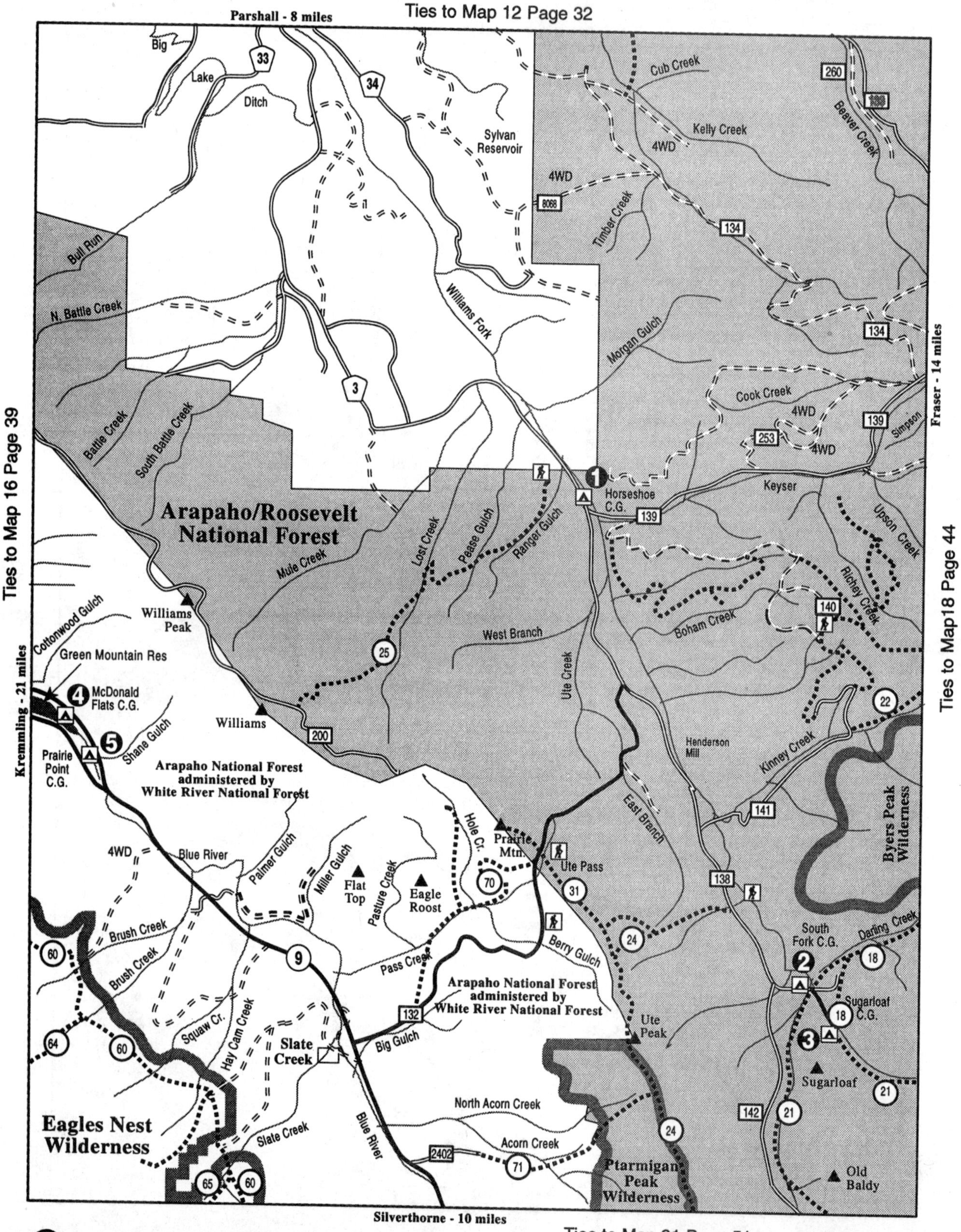

MAP 17

Parshall - 8 miles

Ties to Map 12 Page 32

Big

33

34

Lake

Ditch

Sylvan
Reservoir

260

189

Cub Creek

Kelly Creek

4WD

Beaver Creek

Bull Run

4WD

B068

Timber Creek

134

N. Battle Creek

Williams Fork

Morgan Gulch

134

Ties to Map 16 Page 39

Fraser - 14 miles

3

Battle Creek

South Battle Creek

Cook Creek

4WD

253

139

Simpson

4WD

Keyser

Horseshoe
C.G.

1

139

**Arapaho/Roosevelt
National Forest**

Lost Creek

Pass Gulch

Ranger Gulch

Mule Creek

140

Upson Creek

Richey Creek

Cottonwood Gulch

Williams
Peak

West Branch

Boham Creek

Ties to Map18 Page 44

Green Mountain Res

25

Ute Creek

22

Kremmling - 21 miles

4 McDonald
Flats C.G.

Shane Gulch

Williams

200

Henderson
Mill

Kinney Creek

5

Prairie
Point
C.G.

Byers Peak
Wilderness

141

4WD

Blue River

Palmer Gulch

Miller Gulch

Pasture Creek

Hole Cr.

Prairie
Mtn

East Branch

**Arapaho National Forest
administered by
White River National Forest**

Ute Pass

138

Derting Creek

Brush Creek

60

Flat
Top

Eagle
Roost

70

31

South
Fork C.G.

18

24

2

Brush Creek

9

Pass Creek

Berry Gulch

Sugarloaf
C.G.

18

64

60

Squaw Cr.

Hay Cam Creek

132

**Arapaho National Forest
administered by
White River National Forest**

Ute Peak

3

21

Slate
Creek

Big Gulch

Sugarloaf

142

21

**Eagles Nest
Wilderness**

Slate Creek

Blue River

North Acorn Creek

24

Acorn Creek

65

60

2402

71

**Ptarmigan
Peak
Wilderness**

Old Baldy

Silverthorne - 10 miles

Ties to Map 21 Page 51

110 Trail Number Symbol   2 △ Campground Symbol & Number

Map 17

MAP 17

18 Trail Number	Trail Name	Map Loc.	Distance	Difficulty	Beginning Elev.	Ending Elev.	Ranger District
	Darling Creek	C 10	6.0 mi	More Diff	8,970'	12,130'	Sulphur

ACCESS: Follow County Road 3, 1/2 mile east of Parshall on Highway 40 or take County Road 3, 24 miles south of Kremmling on Highway 9. Follow either to County Road 30 (FDR 138). Continue south up Williams Fork Valley to trailhead before Sugarloaf Campground. **ATTRACTIONS:** The Darling Creek Trail provides access to the Continental Divide by connecting with the St. Louis Divide Trail. The trail is a moderate hike at the start and along Darling Creek. When the trail forks from Darling Creek and follows a tributary, it quickly gains elevation and becomes very steep in sections. When the trail connects with the St. Louis Divide Trail, there is an excellent view of the area. **NARRATIVE:** The Darling Creek Trail begins at Amax's West Portal, a tunnel portal owned by AMAX (American Metals, a division of Climax International Company). The tunnel and train system is used by AMAX to transport molybdenum ore from their mine on the other side of the Continental Divide to their mill site in the Williams Fork Valley.

As the trail approaches and follows Darling Creek, it continues up a moderate slope. While adjacent to the Creek, look for calm pools of water. The bottom will appear to be covered with gold but is actually fools gold. After about 2 miles the trail follows a tributary to the Darling Creek towards the St. Louis Divide. The trail becomes steeper in sections as it gains elevation and the vegetation becomes less dense. Talus slopes on the adjacent hill will frequently come into view. Also visible along this part of the trail is the end of a road built by the Denver Water Board. The final part of the trail reaches timberline and the junction with the St. Louis Divide Trail. **USE:** Medium. **ACTIVITIES:** HIKING, HORSES, FISHING. **USGS:** UTE PEAK, BYERS PEAK QUADS. **MAPS:** 17 & 18.

21 Trail Number	Trail Name	Map Loc.	Distance	Difficulty	Beginning Elev.	Ending Elev.	Ranger District
	South Fork Loop	C 10	27 mi	Difficult	8,970'	8,970'	Sulphur

ACCESS: Follow County Road 3, 1/2 mile east of Parshall on Hwy 40, or take County Road 3, 24 miles south of Kremmling on Hwy 9. Follow either route to County Road 30. Continue up Williams Fork Valley to trailhead before Sugarloaf Campground. **ATTRACTIONS:** This 27 mile circular trail is one of the most popular trails in the Williams Fork Valley. Trail users can start from the trailhead parking lot and follow either the South Fork of the Williams Fork or the Williams Fork River. The trail is suitable for both foot and horse use and many campsites are available. Parts of the upper trail are very steep, however, both the lower trails are moderate hikes. This is a long trail and if users are planning on hiking the entire distance they should be prepared for medical emergencies, altitude sickness, hypothermia and dehydration. **NARRATIVE:** The Williams Fork Trail can be used for day hikes or for overnight trips. The trail follows either the South Fork of the Williams Fork or the Williams Fork River. Trail difficulty varies from easy while following the streams to most difficult where the trail climbs above timberline To begin, the trail users (Hikers, Mtn Bikes.) can start by going through Sugarloaf Campground and crossing the foot bridge over the William Fork or by going adjacent to South Fork Campground and following the South Fork River. Horse access to the trail along Williams Fork River via the Darling Creek Trail.

Following the trail along the South Fork the naturalness of the river area can be seen for the next 10 miles Occasional large old growth Engelmann spruce will be seen along this part of the trail.During spring the snow melt makes the river high and very fast. There are two difficult log bridge crossing in the first 3 miles.

While primarily still in lodgepole pine, the trail begins to gain elevation and occasionally climbs some steep hills. Eventually the vegetation along the trail will be spruce-fir. When higher in elevation, the trail opens into meadows along the river. Views of the Continental Divide can be seen up the valley and the Williams Fork Valley can be seen to the north. As the trail follows the meadows, the valley begins to open up providing a good view of the upper valley. Ptarmigan Pass Trail will drop down from the south and connect with the South Fork Trail. This is where a large elk herd summers and is often seen by trail users.

After following the river for about 10 miles, the trail begins to climb quite steeply. This is the most difficult part of the trail as it climbs to timberline, follows the ridge and then drops down to Bobtail Creek. When following the ridge before dropping down into Bobtail Creek, users should be careful not to drop down into the Middle Fork Valley. This is a steep trail that is no longer maintained.

The trail follows Bobtail Creek for 4 miles, passes Pettingell Peak 13,553 ft. and then comes to the old Bobtail Mine. Located in this area are some old log cabins, some new buildings which belong to the Denver Water Board and the old Jones Pass Wagon Road. From here the trail follows the river below timberline going from spruce-fir back into lodgepole pine and eventually ends up in Sugarloaf Campground. **USE:** Medium. **ACTIVITIES:** HIKING, HORSES, MTN BIKING, FISHING. **USGS:** UTE PEAK SE, DILLON, LOVELAND PASS, BYER PEAKS QUADS. **MAPS:** 17 & 18.

CAMPGROUNDS LOCATED IN MAP 17								
Map No.	Name	Fee	No. of Units	Max. Length	Elev.	Toilets	Water	Ranger District
1.	Horseshoe	$	7	23'	8,540'	Yes	Yes	Sulphur
2.	South Fork	$	21	23'	8,940'	Yes	Yes	Sulphur
3.	Sugarloaf	$	11	23'	8,970'	Yes	Yes	Sulphur

MAP 17

22	Trail Name	Map Loc.	Distance	Difficulty	Beginning Elev.	Ending Elev.	Ranger District
Trail Number	**Kinney Creek**	C 10	4.0 mi	Difficult	9,700'	11,260'	Sulphur

ACCESS: Follow County Road 3, 1/2 mile east of Parshall on High-way 40 or take County Road 3, 24 miles south of Kremmling on Highway 9. Follow either to County Road 30 (FDR 138). Follow County Road 30, 3 miles to Kinney Creek County Road 302. Turn left and continue for 4 miles to the trailhead junction on the right of the road. **ATTRACTIONS:** The trail begins on a gradual slope and follows Kinney Creek through large spruce-fir stands. Within a 0.5 mile, the small valley becomes much narrower and the trail becomes much steeper. At this point camping spots become more limited. The end of the trail provides access to Lake Evelyn, Horseshoe Lake and St. Louis Divide Trails. Ample parking is provided at the trailhead due to two small timber harvest areas in the area. Elk are likely to be seen on the upper part of the trail. **NARRATIVE:** Kinney Creek Trail offers the opportunity to walk through heavy canopies of spruce-fir, open stands with views of the surrounding timberline and cirque areas, and timberline areas with views of the surrounding ridges and valleys.

The trail begins in a flat area and follows the creek through a dense stand of spruce-fir. As the trail climbs in elevation, the valley becomes narrower and steeper. Towards timberline, the stands begin to open with the trees becoming shorter and stunted. At this point the openings give an excellent view of the nearby cirques formed by glaciers. Notice where the glaciers ended and how the narrow V-shaped valley opens into a U-shaped valley. Also in view are Bills Peak, Byers Peak and the adjacent alpine ridges.

Along the upper part of the trail elk will often be seen. The trail ends on a saddle above timberline and provides a view of Lake Evelyn in the next valley to the north, the alpine ridge to the east and the Williams Fork Mountains and Gore Range can be seen to the west. At this point access can be gained to the Lake Evelyn Trail and St. Louis Divide-Jones Pass Trail. **USE:** Medium. **ACTIVITIES:** HIKING, HORSES, FISHING. **USGS:** UTE PEAK, BYERS PEAK QUADS. **MAPS:** 17 & 18.

24	Trail Name	Map Loc.	Distance	Difficulty	Beginning Elev.	Ending Elev.	Ranger District
Trail Number	**Ute Peak**	C 10	15 mi	Difficult	8,890'	11,777'	Sulphur

ACCESS: 24 miles south of Kremmling on Highway 9 to FDR 132, COUNTY RD 3. East on FDR 132 over Ute Pass to FDR 138. South 4 miles to Ute Peak Trailhead. **ATTRACTIONS:** Ute Peak Trail is a 15 mile trail which leaves from the Williams Fork Road (County Road 30) climbs up to the Williams Fork ridge and follows the ridge to Ptarmigan Pass. The trail is steep in parts below the ridge. However, once on the top the trail levels out. Water is not available on the ridge (other than by dropping down off the top) and users should be prepared for windy and cold conditions as well as the high altitude. When on the ridge, excellent views are seen of the Continental Divide. **NARRATIVE:** The Ute Peak Trail is one of the more spectacular trails in the area. Although the hike up to the ridge is steep in parts and strenuous, once on top, panoramic views of the Continental Divide and the Gore range are seen.

The trail begins lodgepole pine and soon passes through a tunnel that goes under train tracks which belong to AMAX (American Metals, a division of Climax International Company). This train system is used by AMAX to transport molybdenum ore from their mine on the other side of the Continental Divide to their mill site in the Williams Fork Valley. The trail follows along a stream and eventually the vegetation begins to open up allowing users to look down on the AMAX mill site.

As the trail gains elevation, it meets with the Ute Pass trail and continues up towards Ute Peak. The climb is steep for 1.5 miles. Lodgepole pine intermingled with Engelmann spruce and sub-alpine fir becomes less dense and shorter. As the trail approaches timberline, occasional views can be seen of the surrounding area. After the trail leaves the trees, it becomes less evident and without treads. From this point the trail basically follows the ridge to Ptarmigan Pass. Wildlife on the ridge is quite abundant. Ptarmigan will often be encountered and large elk herds can often he seen in the upper part of the Williams Fork Valley. **USE:** Moderate. **ACTIVITIES:** HIKING, HORSES, FISHING. **USGS:** UTE PEAK, DILLON QUADS. **MAPS:** 17 & 21.

MAP 17

25 Trail Number	Trail Name	Map Loc.	Distance	Difficulty	Beginning Elev.	Ending Elev.	Ranger District
	Williams Peak	C 9	8.0 mi	Difficult	8,420'	11,180'	Sulphur

ACCESS: 1: 24 miles south of Kremmling on Highway 9, turn left onto Ute Pass County Road 3. Follow 9 miles to junction with FDR 138. Travel north for 3.5 miles to Horseshoe Guard Station on the left. Trail begins behind the Guard Station.
ACCESS 2: 1/2 mile east of Parshall turn onto County Road 3 and follow for 14 miles. Horseshoe Guard Station will be set back on the right with a cattle guard at the entrance to the driveway. Trail begins behind the Guard Station.
ACCESS 3: 12 miles south of Kremmling on Highway 9 turn left onto the Williams Peak County Road 381 (FDR 200). Follow approximately 14 miles to trail junction on left side of road. Not recommended for horse trailers. **ATTRACTIONS:** The first 4 miles of this trail is a moderate hike, however, the last part is quite steep. Beginning on a sagebrush covered hill, the trail quickly crosses into lodgepole pine. Once the trail enters the trees, it crosses onto private land owned by AMAX for about 5 miles until it again reaches the National Forest. There is a right-of-way along the trail while it passes across the private land, so users need not get permission to cross the land. The trail is good for horse use although, while on private land, the horses must be kept on the trail. **NARRATIVE:** The Williams Peak Trail begins behind the Horseshoe Guard Station by going up a small sagebrush covered hill. Lupine can be seen blooming in this area usually around July. In less than 1/2 mile the trail crosses into a lodgepole pine forest which designates the crossing onto private land. A trail right-of-way allows access for 5 miles until the trail crosses onto National Forest lands.

About 2 miles after the trail enters the lodgepole pine, it crosses Pease Gulch. At this point the trail is easily lost and users should be sure to follow tree blazes. After crossing Lost Creek the trail follows the other side through meadows where the trail is easy to confuse with cow paths.

As the trail passes through an old stand of lodgepole pine, it approaches the Forest boundary. The trail passes through a thick stand of old burned over pine and becomes steeper while following a number of switchbacks. It then opens into small sage and aspen parks with a view of the Williams Fork Valley. As the trail approaches FDR 200, it continues steep and fairly open. The trail ends as it meets the Williams Peak Road (FDR 200). **USE:** Medium. **ACTIVITIES:** HIKING, HORSES. **USGS:** SYLVAN RESERVOIR, UTE PEAK, BATTLE MOUNTAIN QUADS. **MAP:** 17.

31 Trail Number	Trail Name	Map Loc.	Distance	Difficulty	Beginning Elev.	Ending Elev.	Ranger District
	Ute Pass	C 10	2.0 mi	Moderate	9,560'	11,060'	Sulphur

ACCESS: 1: 24 miles south of Kremmling on Highway 9, turn left onto Ute Pass County Road 3. Follow 5 1/2 miles to Ute Pass. Trail is on right (east) side of pass.
ACCESS 2: 1/2 mile east of Parshall on Highway 40, take County Road 3, 18 miles to Ute Pass. The road will fork after about 15 miles, stay to right. Trailhead is on (east) side of Ute Pass. **ATTRACTIONS:** Starting at Ute Pass, this 2 mile trail provides an easier access to the higher parts of the Ute Peak Trail than if hikers were to start at the beginning. The trail is suitable for horse and foot travel and is popular with day hikers. Once the trail intersects with Ute Peak Trail, it continues to timberline and follows the Williams Fork Ridge. **NARRATIVE:** The Ute Pass trail is basically a 2 mile tie-in from Ute Pass to the Ute Peak Trail The trial provides access to the Williams Fork Ridge from a higher elevation and therefore makes the hike somewhat easier. The trail is a moderately sloped trail, until it meets the Ute Peak trail where it becomes much steeper.

The trail begins in lodgepole pine and goes through stands of spruce-fir, providing a good opportunity to observe both communities. As the trail connects with the Ute Peak Trail, it begins to open, allowing views of the Gore Range and both the Williams Fork and Blue River Valleys. This trail good for horse travel. **USE:** Medium. **ACTIVITIES:** HIKING, HORSES. **USGS:** UTE PEAK SE QUAD. **MAP:** 17.

MAP 18

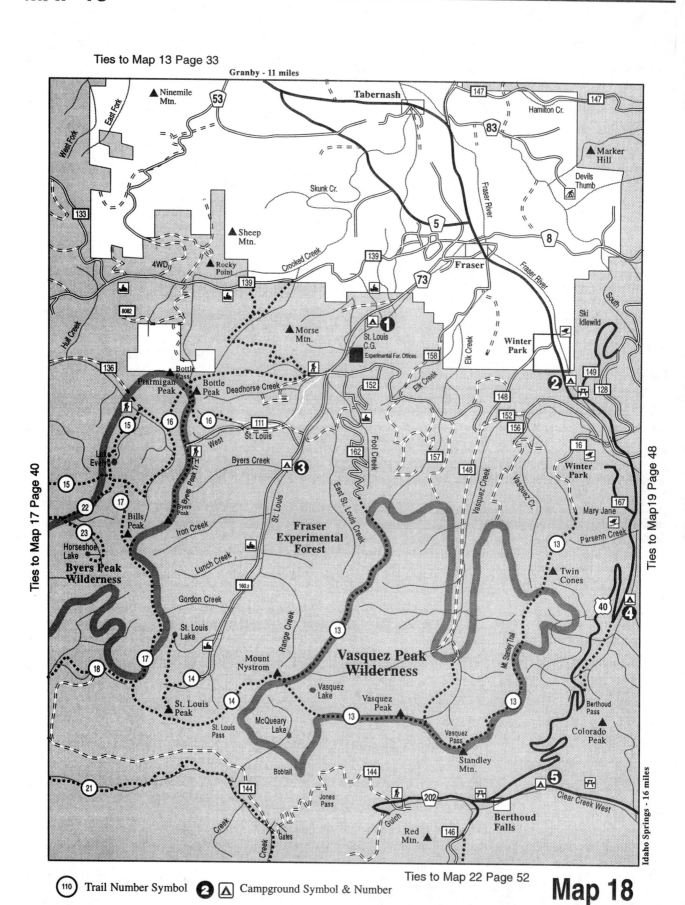

Ties to Map 13 Page 33

Granby - 11 miles

Ties to Map 17 Page 40

Ties to Map19 Page 48

Idaho Springs - 16 miles

Ninemile Mtn.

Tabernash

Hamilton Cr.

Marker Hill

Devils Thumb

Skunk Cr.

West Fork

East Fork

Sheep Mtn.

Fraser River

Crooked Creek

Fraser

Fraser River

4WD

Rocky Point

Hull Creek

Morse Mtn.

St. Louis C.G.

Experimental For. Offices

Winter Park

Ski Idlewild

South

Elk Creek

Bottle Pass

Ptarmigan Peak

Bottle Peak

Deadhorse Creek

St. Louis

West

Byers Creek

Elk Creek

Fool Creek

Vasquez Creek

Vasquez Cr.

Winter Park

Mary Jane

Parsenn Creek

Lake Evelyn

Bills Peak

Byers Peak Tr.

Byers Peak

Iron Creek

Fraser Experimental Forest

East St. Louis Creek

Twin Cones

Horseshoe Lake

Byers Peak Wilderness

Lunch Creek

St. Louis Lake

Gordon Creek

St. Louis Peak

St. Louis Pass

Mount Nystrom

Range Creek

Vasquez Peak Wilderness

Vasquez Lake

Vasquez Peak

Mt. Stanley Trail

Berthoud Pass

Colorado Peak

McQueary Lake

Vasquez Pass

Standley Mtn.

Bobtail

Jones Pass

Gulch

Red Mtn.

Berthoud Falls

Clear Creek West

Gales

Creek

Creek

Creek

Ties to Map 22 Page 52

⑩ **Trail Number Symbol**   ❷ ⬚ Campground Symbol & Number

Map 18

MAP 18

15 Trail Number	Trail Name	Map Loc.	Distance	Difficulty	Beginning Elev.	Ending Elev.	Ranger District
	Lake Evelyn	D 10	5.2 mi	Difficult	9,440'	10,920'	Sulphur

ACCESS: South on County Road 3, 1/2 mile east of Parshall on Highway 40 or take County Road 3, 24 miles south of Kremmling on Highway 9. Follow either to County Road 32 (FDR 139). Turn east onto County Road 32 and follow for 5 miles to fork with FDR 136. Take right fork (FDR 136) 2.5 miles to Lake Evelyn Trailhead. **ATTRACTIONS:** The Lake Evelyn Trail begins on an old logging road which has been closed to motorized vehicles. The road continues for approximately 1.5 miles before turning into a trail. The beginning of the trail is an easy to moderate hike, however, it becomes more difficult as it approaches the lake. Above the lake, the trail is very steep and rocky as it climbs up onto the alpine ridge. Once on the ridge access is provided to either St. Louis Divide or Kinney Creek Trail. The Lake Evelyn Trail continues for another 2 miles on a moderate slope down Keyser Ridge. Good camping sites are available near the lake. **NARRATIVE:** The Lake Evelyn Trail can be accessed from either Keyser Creek (FDR 136) or from Keyser Ridge (FDR 140) with Lake Evelyn basically at the middle of the trail. Approaching from Keyser Creek, the trail starts on an old logging road which has been closed to motorized travel. The trail passes through some old timber sales and across Keyser Creek before it begins to climb. Vegetation on the lower part of the trail is lodgepole pine and as the trail nears the lake, the vegetation changes to spruce-fir. Adjacent to Lake Evelyn are some large Engelmann spruce. Good camping sites are located around the lake and the cutthroat trout around the lake provide good fishing opportunities. Users should be sure to bring fishing poles. From the lake, the trail climbs above timberline and up on the ridge where access can be gained to Kinney Creek Trail and St. Louis Divide Trail. At this point views of the surrounding areas can be seen. From here the trail follows down Keyser Ridge for 2 miles on a moderate slope to the other trailhead. Users who are planning on beginning from Keyser Ridge will need either a truck or four-wheel drive vehicle to drive to the trailhead. **USE:** Medium. **ACTIVITIES:** HIKING, HORSES, FISHING. **USGS:** BOTTLE PASS, BYERS PEAK QUADS. **MAP:** 18.

16 Trail Number	Trail Name	Map Loc.	Distance	Difficulty	Beginning Elev.	Ending Elev.	Ranger District
	Bottle Pass	D 10	1.2 mi	Easy	9,880'	11,390'	Sulphur

ACCESS: Follow County Road 3, 1/2 mile east of Parshall on Highway 40 or take County Road 3, 24 miles south of Kremmling on Highway 9. Follow either to County Road 32 (FDR 139). Turn east onto County Road 32 and follow 5 miles to FDR 136 on right. Take FDR 136, 2 1/2 miles to Lake Evelyn trailhead. Continue on Lake Evelyn trail for 1 mile to junction with Bottle Pass trail. **ATTRACTIONS:** The Bottle Pass trail takes off from the Lake Evelyn trail after approximately 1 mile. The beginning of the trail passes through old logging area and at times is difficult to find, however, once in the trees, it becomes easier to follow. The trail gains elevation as it approaches Bottle Pass and the last half mile is quite steep. **NARRATIVE:** The Bottle Pass trail begins at a fork in an old logging road which the Lake Evelyn Trail follows. At this point the Lake Evelyn trail climbs to the west while the Bottle Pass trail climbs to the east. The beginning of the trail passes through an old logging area where the trail may occasionally be obscured. Once the trail climbs into the trees it begins to climb in elevation. As the trail approaches Bottle Pass, it follows timberline and provides views of Ptarmigan and Bottle Peaks. Users should look for stunted Engelmann spruce near the ridge top to see how they sprawl across the ground and grow only a few inches high. Once up on the pass, views can be seen of Fraser, the Williams Fork Valley and the Continental Divide. **USE:** Light. **ACTIVITIES:** HIKING, HORSES. **USGS:** BOTTLE PASS QUAD. **MAP:** 18.

Map No.	Name	Fee	No. of Units	Max. Length	Elev.	Toilets	Water	Ranger District
CAMPGROUNDS LOCATED IN MAP 2								
1.	St. Louis Creek	$	18	32'	9,000'	Yes	Yes	Sulphur
2.	Idelwild	$	26	32'	9,000'	Yes	Yes	Sulphur
3.	Byers Creek	$	6	32'	9,400'	Yes	Yes	Sulphur
4.	Robbers Roost	$	11	45'	9,826'	Yes	Yes	Sulphur
5.	Mizpah	$	10	20'	9,600'	Yes	Yes	Clear Creek

MAP 18

17 Trail Number	_Trail Name	Map Loc.	Distance	Difficulty	Beginning Elev.	Ending Elev.	Ranger District
	St Louis Divide-Jones Pass	D 10	16 mi	Moderate	10,080'	12,450'	Sulphur

ACCESS: Follow County Road 3, 0.5 mile east of Parshall on Highway 40 or take County Road 3, 24 miles south of Kremmling on Highway 9. Follow either to County Road 32 (FDR 139). Turn east onto County Road 32 and follow 5 miles to FDR 136 on right. Take FDR 136 2.5 miles to Lake Evelyn trailhead. Continue on Lake Evelyn and Bottle Pass trails to St. Louis Creek and Darling Creek trails. **ACCESS 2:** Via Lake Evelyn, Kinney Creek and Darling Creek trails.

ATTRACTIONS: The St. Louis Divide-Jones Pass trail begins a short way up the Bottle Pass trail. After forking from the Bottle Pass trail, the St. Louis Divide-Jones Pass trail follows upper Keyser Creek to just below the creeks head lakes. At this point the trail begins to climb up on to the divide, where the trail follows the open alpine ridge to Jones Pass. The majority of the trail is a moderate with occasional difficult steep sections. **NARRATIVE:** The St. Louis Divide-Jones Pass trail is mostly a moderate hike which allows access to the alpine ridge without an extremely difficult hike. The trail takes off of the Bottle Pass trail and follows Keyser Creek towards its headwaters. The trail then begins climbing the ridge towards timberline. After 2 1/2 miles the trail meets the junction with Kinney Creek trail above the Lake Evelyn trail.

Once above the timberline, the trail follows the ridge to the south. The trail is not maintained and although times there is no defined trail to follow, it basically follows the top along side of the ridge. Trail users should have no difficulties following the ridge. From the ridge top, excellent views of mountain ridges can be seen in most directions including part of the Continental Divide, the Gore Range and the Williams Fork Mountains. In the area surrounding the ridge, alpine lakes valleys, and streams can be seen providing a good opportunity to observe the alpine community. The elk herds which summer in the upper Williams Fork can often be seen from the ridge. As the wildflowers begin to emerge in early summer, the alpine becomes very colorful

As the trail follows the ridge some of the major peaks it views are Bill's Peak, St. Louis Peak, and Mt. Nystrom. Trails joining this trail along the ridge include the Darling Creek, St. Louis and Jones Pass Trails The last 2 miles of the trail follow the Continental Divide. Once Jones Pass is reached, access is on either the east or west side of the divide. **USE:** Light. **ACTIVITES:** HIKING, HORSES, FISHING. **USGS:** UTE PEAK SE, BYERS PEAK QUADS. **MAP:** 18.

Iceberg Lake -- Fischer USFS

Continued From Page 37

unless there is more danger in remaining where you are; use extreme care in moving the injured.

Altitude Sickness

A person should spend 2 or 3 days getting acclimatized to high altitudes before hiking. The lack of oxygen at high elevations gives some travelers altitude sickness.

Prevention: The best prevention is slow ascent with gradual acclimatization to altitude. Beginning at an elevation of 9,000 feet, it is recommended that you do not ascend more than 1,000 vertical feet per day.

Symptoms:

a. Cough

b. Lack of appetite;

c. Nausea or vomiting;

d. Staggering gait; and

e. Severe headaches.

Treatment: A person with symptoms of altitude sickness should breathe deeply, rest and eat quick-energy foods such as dried fruit or candy. Take aspirin to help the headaches; antacid pills may help other symptoms. If symptoms persist, seek lower elevations immediately. Continued exposure can make the victim too weak to travel and may lead to serious complications.

Dehydration

Adults require 2 quarts of water daily and up to 4 quarts for strenuous activity at high elevations. There is a 25 percent loss of stamina when an adult loses 1 1/2 quarts of water. To avoid dehydration, simply drink water as often as you feel thirsty. The "don't drink" when hiking saying is nonsense. An excellent way to determine if you are becoming dehydrated is to check your urine; dark yellow urine may indicate you are not drinking enough water.

Hypothermia

Be aware of the danger of hypothermia subnormal temperature of the body. Lowering of internal temperature may lead to mental and physical collapse. Hypothermia is caused by exposure to cold, and it is aggravated by wetness, wind, and exhaustion. It is the number one killer of outdoor recreationists.

Cold Kills in Two Distinct Steps

1. The first step is exposure and exhaustion. The moment you begin to lose heat faster than your body produces it, you are undergoing exposure. Two things happen: you voluntarily exercise to stay warm and your body makes involuntary adjustments to preserve normal temperature in the vital organs. Both responses drain your energy reserves. The only way to stop the drain is to reduce the degree of exposure.

2. The second step is hypothermia. If exposure continues until your energy reserves are exhausted, cold reaches the brain, depriving you of judgment and reasoning power. You will not be aware that this is happening: You will lose control of your hands. This is hypothermia. Your internal temperature is sliding downward. Without treatment, this slide leads to stupor, collapse and death.

Defense Against Hypothermia

Stay dry. When clothes get wet, they lose about 90 percent of their insulating value. Wool loses less heat than cotton, down, and some other synthetics. Choose rain clothes that cover the head, neck, body, and legs, and provide good protection against wind-driven rain. Polyurethane coated nylon is best. The coatings won't last forever.

Understand cold. Most hypothermia cases develop in air temperatures between 30 and 50 degrees.

Symptoms: If you or a member of your party is exposed to wind, cold and wet, think hypothermia. Watch yourself and others for these symptoms:

a. Uncontrollable fits of shivering. Vague, slow, slurred speech.

b. Memory lapses, incoherence.

c. Immobile, fumbling hands.

d. Frequent stumbling, lurching gait. Drowsiness - to sleep is to die.

e. Apparent exhaustion. Inability to get up after a rest.

Treatment

a. The victim may deny any problem. Believe the symptoms, not the victim. Even mild symptoms demand immediate treatment. Get the victim out of the wind and rain. Strip off all wet clothes.

b. If the victim is only mildly impaired, give warm drinks. Get the person into warm clothes and a warm sleeping bag. Well wrapped, warm (not hot) rocks or canteens will help.

c. If victim is badly impaired, attempt to keep him/ her awake. Put the victim in a sleeping bag with another person - - both stripped. If you have a double put the victim between two warm people. Build a fire to warm the camp.

If You Get Lost

Someone in your party may become lost. If you or someone else becomes lost, follow these steps.

a. Stay calm and try to remember how you got to your present location. Look for familiar land marks, trails or streams. If you are injured, near exhaustion or its dark, stay where you are; some one may be looking for you. If you decide to go on, do it slowly.

b. Try to find a high point with a good view and then plan your route of travel. When you find a trail, stay on it. If you stay lost, follow a drainage down stream. In most cases it will eventually bring you to a trail or to a road. Help won't be far off.

c. When backpacking with children, be sure they stay with you or near camp. Discuss with them what they are to do if they become separated. They should know the international distress signals and when to use them - three smokes, three blasts on a whistle, three shouts, three flashes of light, three of anything that will attract attention.

d. A guaranteed method of attracting attention and getting someone to investigate during the summer months is a fire creating large volume of smoke. Green boughs on fire will create smoke. A fire should only be used as a last resort. Be sure your fire does not escape and cause a wildfire. You can be held liable for the entire cost of putting it out!

What To Do When Someone is Overdue

Stay calm and notify the County Sheriff or Ranger in the trip area. They will take steps to alert or activate a local search and rescue organization. If the missing person returns later, be sure to advise the Sheriff or Ranger.

Backpacking article courtesy of the U.S. Forest Service with credit to:

Bureau of Land Management

Colorado Horsemen's Council

Colorado Open Space Council

Colorado Outward Bound School

Izaak Walton League Of America

National Leadership School

Sierra Club

Wilderness Society

MAP 19

Ties to Map 14 Page 34

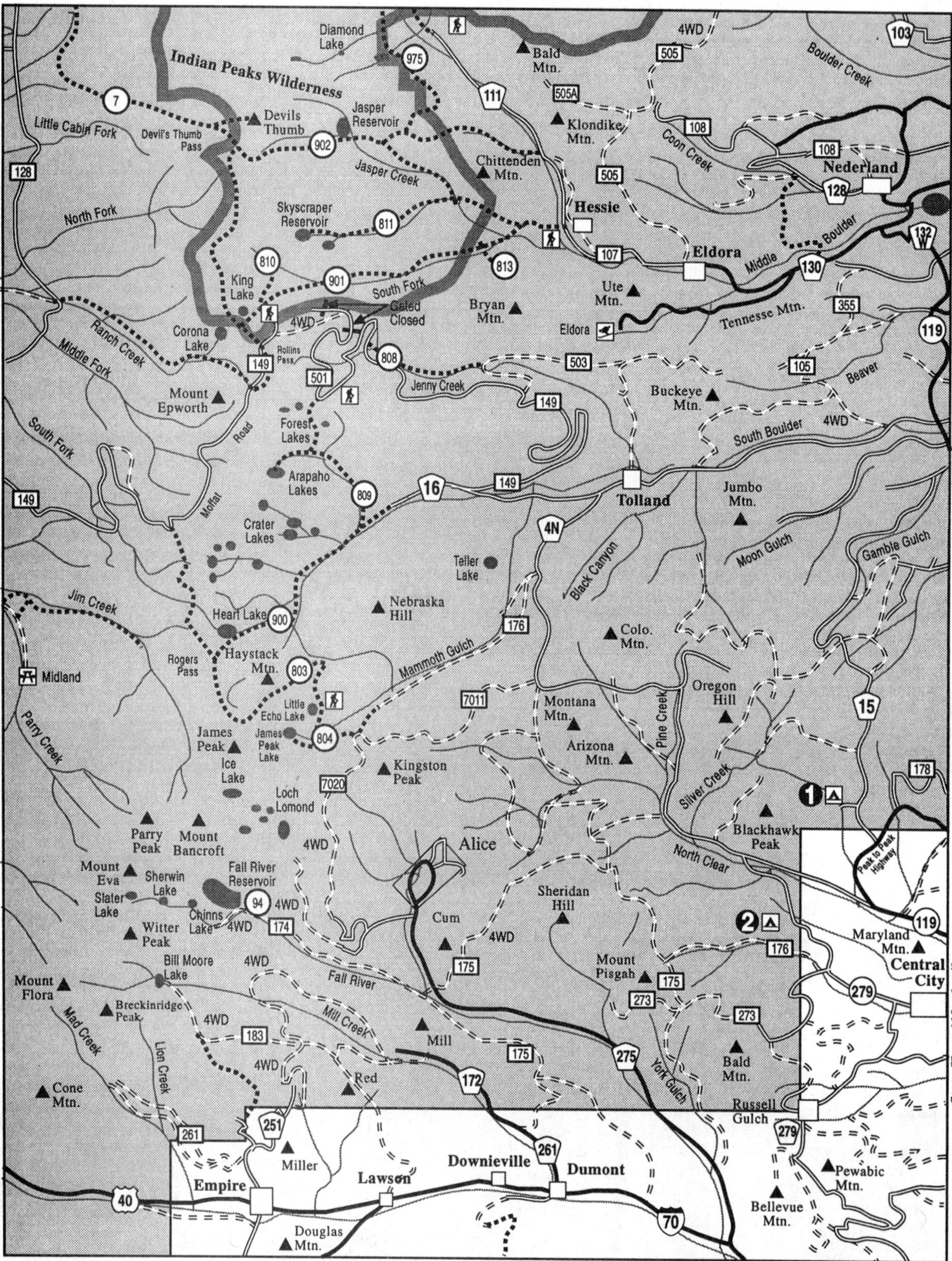

Boulder - 16 miles

Diamond Lake
Bald Mtn.
4WD
Boulder Creek
103
975
505
111
505A
Indian Peaks Wilderness
Klondike Mtn.
108
7
Jasper Reservoir
Chittenden Mtn.
Coon Creek
108
Nederland
Little Cabin Fork
Devils Thumb
902
505
128
Devil's Thumb Pass
Hessie
Fraser - 4 miles
128
Jasper Creek
107
Eldora
Middle Boulder
132
North Fork
Skyscraper Reservoir
811
813
130
810
901
King Lake
South Fork Gated Closed
Bryan Mtn.
Ute Mtn.
Tennesse Mtn.
355
119
Ranch Creek
Corona Lake
Eldora
Middle Fork
149
Rollins Pass
4WD
808
503
Buckeye Mtn.
105
Beaver
Mount Epworth
501
Jenny Creek
149
South Boulder
4WD
Winter Park - 5 miles
South Fork
Road
Forest Lakes
149
149
Jumbo Mtn.
Moffat
Arapaho Lakes
809
16
Tolland
Gamble Gulch
Moon Gulch
4N
Crater Lakes
Teller Lake
Black Canyon
Ties to Map 20 Page 50
Jim Creek
Heart Lake
900
Nebraska Hill
176
Colo. Mtn.
Oregon Hill
15
Rogers Pass
Haystack Mtn.
803
7011
Montana Mtn.
Pine Creek
Silver Creek
178
Midland
Little Echo Lake
804
Arizona Mtn.
Parry Creek
James Peak
James Peak Lake
7011
Kingston Peak
1
Loch Lomond
7020
Blackhawk Peak
North Clear
Parry Peak
Mount Bancroft
4WD
Alice
Peak to Peak Highway
Mount Eva
Sherwin Lake
Fall River Reservoir
Sheridan Hill
2
Maryland Mtn.
Slater Lake
Chinns Lake
94
4WD
Cum
4WD
176
Central City
Witter Peak
174
175
Mount Pisgah
175
279
119
Bill Moore Lake
4WD
Fall River
273
273
Mount Flora
Breckinridge Peak
4WD
Mill Creek
Mill
175
Bald Mtn.
Lion Creek
183
4WD
Red
275
Russell Gulch
Mad Creek
Cone Mtn.
172
279
Berthoud Falls - 7 miles
261
251
Pewabic Mtn.
Empire
Miller
Downieville
261
Dumont
Bellevue Mtn.
40
Douglas Mtn.
Lawson
70

Georgetown - 4 miles

Idaho Springs - 4 miles

Nederland - 18 miles
State Hwy. 119 - 1 mile

Ties to Map 18 Page 44

110 Trail Number Symbol
2 △ Campground Symbol & Number
Ties to Map 23 Page 54

Map 19

MAP 19

803/804 Trail Number	Trail Name	Map Loc.	Distance	Difficulty	Beginning Elev.	Ending Elev.	Ranger District
	Ute & James Peak Lake	F 10	*	Moderate	10,400'	11,860'	Boulder

ACCESS: Take either Colorado 119 or 72 to Rollinsville. Turn west onto the Rollins Pass Road about 6 miles to Tolland. Take the road to your left marked "Apex, Mammoth Gulch. " The road is fairly rough, becoming a four-wheel drive route after 1.5 miles. Park here and hike 3.5 miles to the trail head or drive on the designated four-wheel drive route. **ATTRACTIONS:** This trail drops gradually to James Peak Lake, set in the basin below the east face of James Peak. Fishing, picnicking, and camping are the principal attractions. You may wish to hike up James Peak from either the trailhead to James Peak Lake, following the south ridge, or by taking the Ute Trail #803 which intersects the James Peak Lake trail approximately 0.5 miles from the trailhead. The Ute Trail #803 heads north past Little Echo Lake, then climbs a ridge and heads west on the ridgetop to the Divide. From here, hikers may walk up the north ridge of James Peak or hike north along the Divide to Rogers Pass, where Rogers Pass Trail drops into the steep cirques of South Boulder Creek drainage. The view from the Divide and James Peak is expansive, while James Peak Lake offers a classic alpine setting. **USE:** Moderate and Heavy. **ACTIVITIES:** HIKING, MTN BIKING, HORSES, FISHING. **USGS:** EMPIRE QUAD. **MAP:** 19.

* TRAIL DISTANCE:

803	UTE:	3.0 MILES.
804	JAMES PEAK LAKE:	1.0 MILES.

810/901 Trail Number	Trail Name	Map Loc.	Distance	Difficulty	Beginning Elev.	Ending Elev.	Ranger District
	Bob, Betty & King Lakes	G 9	*	Mod/Diff	9,100'	11,600'	Boulder

ACCESS 1: Hessie Townsite and west of Eldora, County Road 130. **ACCESS 2:** Foot trail north of Rollins Pass intersects a trail to King Lake and Devils Thumb Pass. **ATTRACTIONS:** The route starts on private lands. Please respect it and stay on the trail. One enters the wilderness at 1.5 miles. Fishing, camping, horse use and hiking available. You may hike from King Lake up another 300 vertical feet to Rollins Pass. Please adhere to the posted regulations if you enter the wilderness. **NARRATIVE:** Both trails follow the same route from Hessie on an old road to a trail leading through montane forests to these alpine lakes, passing below the Moffat Road Railroad over Rollins Pass and up steep meadowed basins in the last mile. Trails to Devils Thumb and Lost Lake are intersected along the route at the lower end where signs point out an obvious fork. **USE:** Moderate. **ACTIVITIES:** HIKING, HORSES, FISHING. **USGS:** EAST PORTAL, NEDERLAND QUADS. **MAP:** 19.

* TRAIL DISTANCE:

810	KING LAKE:	5.0 MILES.
901	BOB & BETTY LAKES	5.5 MILES.

CAMPGROUNDS LOCATED IN MAP 19

Map No.	Name	Fee	No. of Units	Max. Length	Elev.	Toilets	Water	Ranger District
1.	Pickle Gulch - Group	$			9,100'	Yes	Yes	Clear Creek
2.	Columbine	$	47	20'	9,020'	Yes	Yes	Clear Creek

Granby Reservoir
W. W. Walker USFS

MAP 20

No trail descriptions for this map

CAMPGROUNDS LOCATED IN MAP 2

Map No.	Name	Fee	No. of Units	Max. Length	Elev.	Toilets	Water	Ranger District
1.	Kelly Dahl	$	46	40'	8,600'	Yes	Yes	Boulder
2.	Cold Springs	$	38	50'	9,200'	Yes	Yes	Clear Creek

Ties to Map 15 Page 38

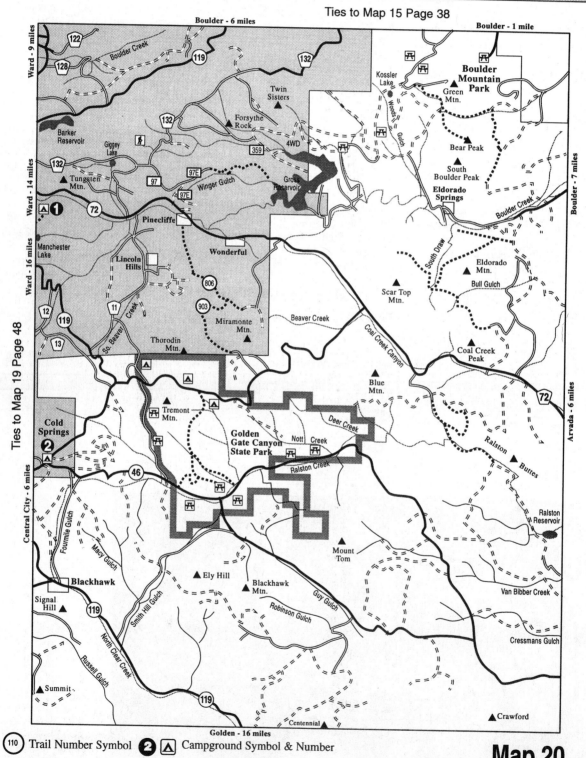

 Trail Number Symbol   ❷ ⧊ Campground Symbol & Number

Map 20

MAP 21

No trail descriptions for this map

No campgrounds located in Arapaho Roosevelt National Forest on this map

Ties to Map 17 Page 40

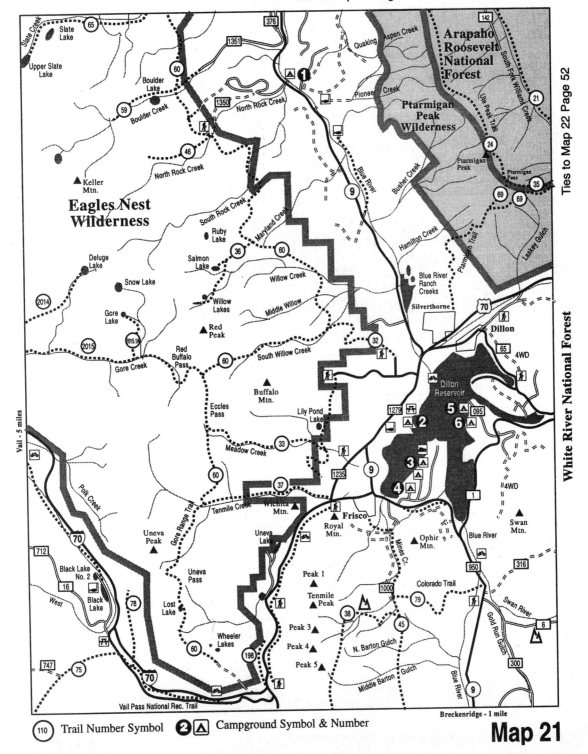

⑴⑩ Trail Number Symbol   ❷△ Campground Symbol & Number

Breckenridge - 1 mile

Map 21

MAP 22

No campgrounds located on this map

Ties to Map 18 Page 44

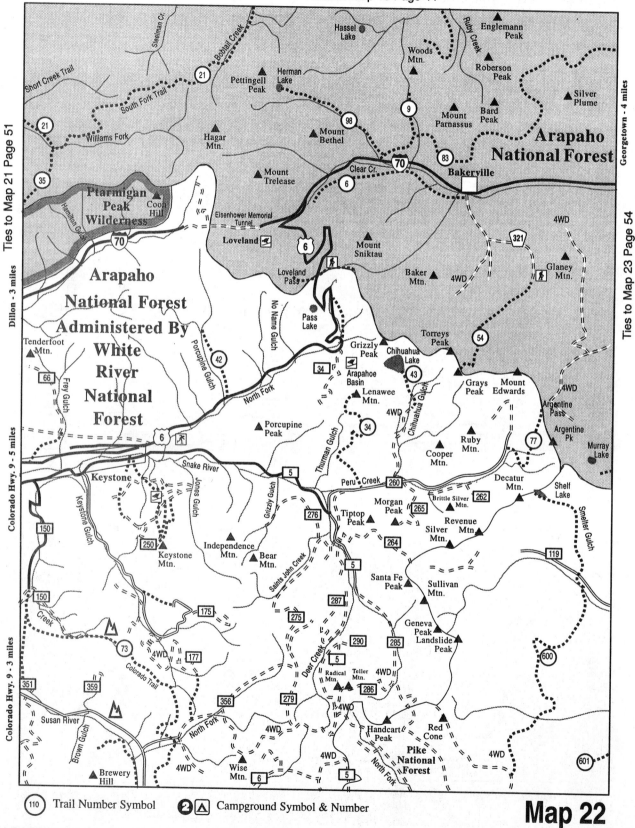

Ties to Map 21 Page 51

Dillon - 3 miles

Colorado Hwy. 9 - 5 miles

Colorado Hwy. 9 - 3 miles

Georgetown - 4 miles

Ties to Map 23 Page 54

Arapaho National Forest

Arapaho National Forest Administered By White River National Forest

Ptarmigan Peak Wilderness

Pike National Forest

(110) Trail Number Symbol   ❷▲ Campground Symbol & Number

Map 22

MAP 22

35 Trail Number	Trail Name	Map Loc.	Distance	Difficulty	Beginning Elev.	Ending Elev.	Ranger District
	Ptarmigan Pass	D 11	3.0 mi	Difficult	10,220'	11,777'	Sulphur

ACCESS: 1: 8 miles up the South Fork Trail along the South Fork of the Williams Fork. **ACCESS: 2:** Take Ute Peak Trail to Ptarmigan Pass. **ATTRACTIONS:** The Ptarmigan Pass Trail is a steep 3 mile trail which connects the South Fork Trail to the end of the Ute Peak Trail. The trail is a difficult hike especially in the beginning. Once on the Pass, access is provided north along the Williams Fork Ridge and excellent views of the area can be seen. Elk are often seen near the trail and throughout the upper Williams Fork Valley. **NARRATIVE:** The beginning of Ptarmigan Pass Trail provides an excellent view looking down the Williams Fork Valley and up towards the Continental Divide. As the trail gains elevation even more spectacular views can be seen.

The trail branches from the South Fork and quickly begins to climb out of the valley. For about 2 miles the trail follows a number of steep switchbacks and then breaks into a stand of trees. At this point the trail follows along the side of the ridge and begins to level out somewhat before reaching the top of the Pass. Once on top, excellent views can be seen of the Continental Divide, the Williams Fork Valley, the Blue River Valley and the Gore Range. Junction with the Ute Peak Trail allows access north along the Williams Fork Ridge.

The entire upper Williams Fork Valley is elk country and since the Ptarmigan Pass Trail is open with good views of the upper valley, elk herds are very often seen. Users may also see elk at the upper section of the trail. **USE:** Light. **ACTIVITIES:** HIKING. **USGS:** DILLON, LOVELAND PASS QUADS. **MAPS:** 21 & 22.

54 Trail Number	Trail Name	Map Loc.	Distance	Difficulty	Beginning Elev.	Ending Elev.	Ranger District
	Grays Peak	E 11	2.5 mi	More Diff	11,200'	14,270'	Clear Creek

ACCESS: I-70 west of Idaho Springs, take the Bakerville exit, travel Stevens Gulch Road four miles. **ATTRACTIONS:** One can climb two 14,000 ft. peaks in one day, Grays and Torreys, getting spectacular views of both sides of the Divide. Grays is one of the easiest 14,000 ft. peaks to climb in the State of Colorado. **NARRATIVE:** The Grays Peak Trail offers a chance to hike above timberline to two 14,000 ft. peaks. There are excellent opportunities to photograph the surrounding scenery on both sides of the Continental Divide. Even in the summer months snowfields remain at this high elevation. The trail is basically for day use. With camping possible above the trailhead. It is a good, fairly easy climb, but heavily used on weekends. Forest Service regulations prohibit the use of motor vehicles on this trail. **USE:** Heavy. **ACTIVITIES:** HIKING, MTN BIKING, HORSES. **USGS:** GRAYS PEAK QUAD. **MAP:** 22.

Corona Range -- G. Lloyd USFS

MAP 23

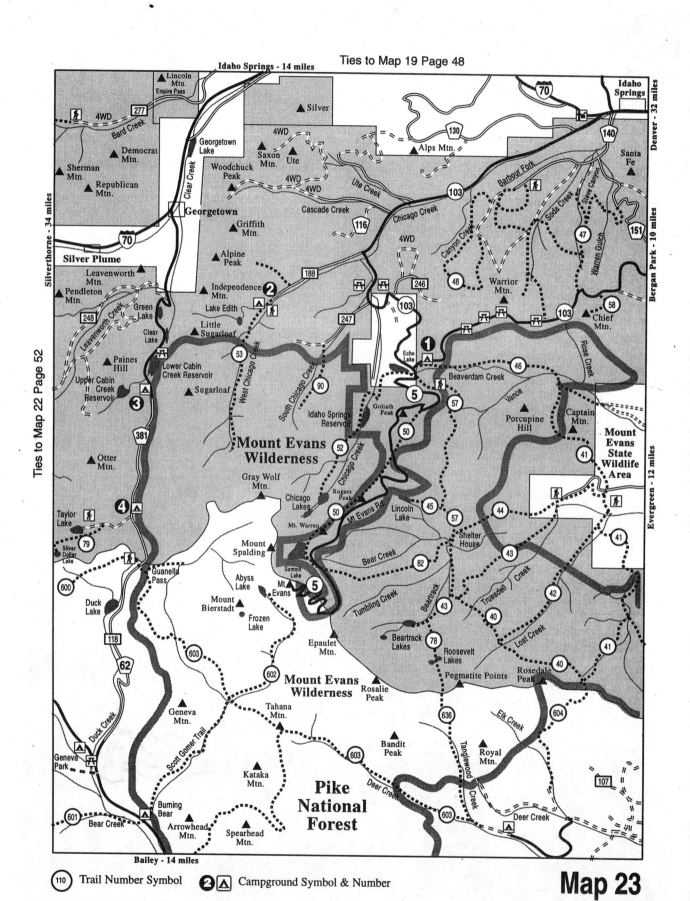

Ties to Map 19 Page 48

MAP 23

43	Trail Name	Map Loc.	Distance	Difficulty	Beginning Elev.	Ending Elev.	Ranger District
Trail Number	**Beartracks**	G 12	6.0 mi	Moderate	9,100'	11,090'	Clear Creek

ACCESS: From Evergreen take the Upper Creek Road for ten miles to the Mount Evans Elk Management Area. This road is closed intermittently in the fall for hunting season and from January 1 to June 15. **ATTRACTIONS:** The trail begins in the forest and travels through an old burn. The wildflowers and surrounding scenery dominate the landscape forming an attractive contrast to the stark, bare trees of the burn. As true of most trails in the Bear Creek Basin, it is possible to reach other trails off the Beartracks Trail. Beartrack Lakes at the end of the trail offer fishing and many good campsites. **NARRATIVE:** Beginning at Camp Rock Trailhead, the trail travels through a forested environment along a creek for a mile or so. At the Mount Evans Wilderness boundary it starts gradually climbing upwards for several miles the scene then opens up as the trail passes through an old burn. This offers an opportunity to view the force of Mother Nature in the stark burned trees in contrast with the subtle beauty of the wildflowers and surrounding peaks. The trail then gets back into the forest before arriving at a beautiful lake nestled at the base of a spectacular rock cliff. The area offers opportunities for camping, hiking, fishing, rock climbing, horseback riding and photography. **USE:** Heavy. **ACTIVITIES:** HIKING, HORSES, FISHING. **USGS:** HARRIS PARK QUAD. **MAP:** 23.

44	Trail Name	Map Loc.	Distance	Difficulty	Beginning Elev.	Ending Elev.	Ranger District
Trail Number	**Beavers Meadows**	G 12	7.0 mi	Moderate	9,050'	11,000'	Clear Creek

ACCESS: Take Hwy 74 to Evergreen, west on Upper Bear Creek Road for 10 miles to Mt. Evans Elk Management Area, follow road 5 miles to Camp Rock Trailhead. The elk management area is closed from January 1 through June 15. **ATTRACTIONS:** Several beaver ponds are easily accessible in about 1 mile on this trail to the Mount Evans Wilderness boundary. Crossing through several meadows the opportunity is great for viewing a variety of wildflowers. From this trail it is possible to reach Beartrack Lakes, Lincoln Lake, Echo Lake or Meridian Pass. **NARRATIVE:** The trail follows an old fire road for about a mile to a scenic meadow with a variety of wildflowers. The meadows are dotted with several beaver ponds which are good for fishing or a lunch spot. Continuing on, the trail climbs to another meadow, Rest House Meadows. From here it is possible to get to Lincoln Lake and Echo Lake. The Beaver Meadows Trail leaves the meadow and climbs through an old burn area. The bark stripped trees provide an interesting contrast against the distant peaks. The trail ends at the junction to Beartrack Lakes and Meridian Pass. **USE:** Heavy. **ACTIVITIES:** HIKING, HORSES, FISHING. **USGS:** HARRIS PARK QUAD. **MAP:** 23.

57	Trail Name	Map Loc.	Distance	Difficulty	Beginning Elev.	Ending Elev.	Ranger District
Trail Number	**Resthouse Meadows**	G 11	6.0 mi	Moderate	10,600'	10,400'	Clear Creek

ACCESS: Highway 103, 14 miles south of Idaho Springs at Echo Lake Campground. **ATTRACTIONS:** The trail is a good trail except for the last mile which is fairly steep, narrow, and rocky. After 1 1/2 miles traveling through the forest, the trail crosses Vance Creek which provides a good picnic or camp site. The trail continues on through an old burn. This affords good views of the surrounding mountains. The charred trees contrast with flowers underneath. The meadow has many wildflowers and two beaver ponds. Deer may be seen along the trail and elk in the meadows. One can also get to Lincoln Lake, Beaver Meadows, and Beartrack Lakes from this trail. **NARRATIVE:** The trail offers a chance for a hike through the woods with easy access. It travels gradually uphill, then drops down to Vance Creek, a nice spot to camp or picnic. The trail continues on up through an old burn, offering a good view of surrounding scenery, then drops down fairly steeply into Resthouse Meadows. The meadow has several beaver ponds and fields of wildflowers. Of special attraction in the meadow is the elephant head flower, in the summer. **USE:** Heavy on weekends. **ACTIVITIES:** HIKING, HORSES. **USGS:** IDAHO SPRINGS, HARRIS PARK QUADS. **MAP:** 23.

CAMPGROUNDS LOCATED IN MAP 23								
Map No.	Name	Fee	No. of Units	Max. Length	Elev.	Toilets	Water	Ranger District
1.	Echo Lake	$	18	20'	10,600'	Yes	Yes	Clear Creek
2.	West Chicago Creek	$	16	30'	9,600'	Yes	Yes	Clear Creek
3.	Clear Lake	$	8	15'	10,000'	Yes	Yes	Clear Creek
4.	Guanella Pass	$	18	35'	10,900'	Yes	Yes	Clear Creek

MAP 23

Trail Number 58	Trail Name	Map Loc.	Distance	Difficulty	Beginning Elev.	Ending Elev.	Ranger District
	Chief Mountain	G 11	1.0 mi	Easy	10,800'	11,700'	Clear Creek

ACCESS: On Highway 103 five miles past Echo Lake towards Bergen Park. **ATTRACTIONS:** This trail is an easily accessible short hike to a beautiful view. From the top of Chief Mountain to the south you overlook the Bear Creek Basin including such peaks as Mt. Evans, Mt. Goliath, Rogers Peak, and Rosalie Peak. **NARRATIVE:** Chief Mountain Trail is an easily accessible short hike to a mountain summit. The trail passes through a spruce-fir forest, then climbs above timberline to the alpine tundra for the last half mile. The last stretch of the trail provides scenic views of the surrounding area. **USE:** Moderate. **ACTIVITIES:** HIKING, MTN BIKING, HORSES. **USGS:** IDAHO SPRINGS QUAD. **MAP:** 23.

Trail Number 79	Trail Name	Map Loc.	Distance	Difficulty	Beginning Elev.	Ending Elev.	Ranger District
	Silver Dollar Lake	F 12	1.5 mi	Moderate	11,000'	12,200'	Clear Creek

ACCESS: I-70 west to Georgetown up the Guanella Pass Road to the first dirt road after the Guanella Pass Campground. **ATTRACTIONS:** This scenic trail offers a short hike to two timberline lakes surrounded by alpine tundra. Fishing is done in both lakes as well as small ponds surrounding the upper lake. The trail overlooks private property around Naylor Lake. Please respect private property. **NARRATIVE:** This is a fairly easy, short hike that gets a bit steep. It climbs above timberline to Silver Dollar Lake. Since the lake is above timberline it is easy to bushwhack around to various peaks and lakes such as Square Top Mountain and Murray Lake. The view from Square Top Mountain over the Continental Divide is also a treat. **USE:** Moderate. **ACTIVITIES:** HIKING, MTN BIKING, HORSES, FISHING. **USGS:** MT. EVANS, MONTEZUMA QUADS. **MAP:** 23.

Green Lake -- G. Lloyd USFS

Rocky Mountain National Park
by National Park Service

Rocky Mountain National Park is located in north central Colorado. From the east it can be reached by automobile on U.S. 34, U.S. 36 and Colorado 7, and from the west by U.S. 40. The nearest major rail, air and bus terminals are at Denver, 65 miles from Estes Park and at Cheyenne, 91 miles to the northeast.

The snow mantled peaks of Rocky Mountain National Park rise above verdant subalpine valleys and glistening lakes. One third of the park is above tree line and here alpine tundra predominates, a major reason these peaks and valleys have been set aside as a national park. This area was first traversed by French fur traders. In 1859 Joel Estes and his son, Milton, rode into the valley that bears their name. Few others settled in this rugged country. About 1909 Enos Mills, a naturalist, writer and conservationist began to campaign for preservation of this pristine area. Mill's campaign succeeded and the area became Rocky Mountain National Park in 1915. A feature of the park is the marked differences found with the changing elevation. At lower levels in the foothills and montane life zone, open stands of ponderosa pine and juniper grow on the slopes facing the sun, on cooler north slopes are Douglas-fir. Gracing the streamsides are blue spruces intermixed with dense stands of lodgepole pine. Here and there appear groves of aspen. Wildflowers dot meadows and glades. Above 9,000 feet forests of Englemann spruce and subalpine fir take over in the subalpine life zone. Openings in these cool, dark forests produce wildflower gardens of rare beauty and luxuriance, where the blue Colorado columbine reigns. At the upper edges of this zone, the trees are twisted, grotesque and hug the ground. Then the trees disappear and you are in the alpine tundra a harsh, fragile world. Here, more than one-quarter of the plants you will see can also be found in the Arctic. From the valleys to its mountain tops, Rocky Mountain National Park encompasses many worlds. In this section are described 29 trails within the Park, we invite you to explore them.

Accommodations
There are no motels or hotels in the park.

Camping
Camping limited park wide to 7 days, June through September. Longs Peak Campground (tents only) has a 3 day limit. The campgrounds fill early each day in the summer. There are no showers or recreation vehicle connections in any campground. Dump stations are at Moraine Park, Glacier Basin, and Timber Creek. Telephones are at Moraine Park, Glacier Basin, Timber Creek and Aspenglen. Wood fires are permitted only in fire grates in the campgrounds and picnic areas. Permits for fires outside these areas are required. Wood gathering is prohibited, firewood bundles are sold at the campgrounds. Pets are permitted in the campgrounds on a leash shorter than 6 feet. (2 meters) Complete services are available at Estes Park, east of the park and at Grand Lake to the West. Reservations for Moraine Park and Glacier Basin family camping, and Glacier Basin Group Areas are available during the summer. Reservations can be made as early as 8 weeks in advance. For backcountry camping and bivouac climbing permits for information, call (970) 586-1206.

Regulations
A permit is required for all overnight stays in the backcountry. The free permits may be obtained in advance or upon arrival at park headquarters, the West Unit Office and at most ranger stations. Backcountry camping is limited to seven nights between May and September and 15 nights during the remainder of the year. Backcountry camping is allowed in designated campsites only or cross country zones by permit only. No pets are permitted in the backcountry.

Fishing
In the mountain streams and lakes of Rocky Mountain National Park are four species of trout: German brown, rainbow, brook and cutthroat. These cold waters may not produce large fish but you will enjoy the superb mountain scenery as you fish.

Remember, you must have a valid Colorado fishing license. Use of live bait is prohibited except under certain special conditions. Review the special fishing regulations at park headquarters or at the nearest park ranger station before you fish. Fishing is not permitted in Bear Lake at any time. Other lakes and streams in the park are under restrictions to protect the Colorado River greenback cutthroat that is being reintroduced to its native habitat. Check with a ranger for details.

Climbing
For the climber, Rocky Mountain National Park offers a variety of challenging ascents throughout the year. A park concessionaire operates a technical climbing school and guide service that provides climbing and mountaineering instruction. For more information, contact park headquarters.

It is important to be familiar with the park's climbing regulations before you begin. These regulations have been established to provide as safe and satisfactory a situation for climbers as possible. Study them and check with a park ranger if you have any questions. Permits are not required for day hiking.

Technical climbs involving only day long excursions do not require registration either at the trailhead or in advance, but registration is always required for overnight bivouacs. It is your responsibility to leave details about your destination with someone who can report your absence if you happen to be overdue.

General
Hiking season in the Park is normally from Mid-June through September. Elevation, location and snow conditions dictate when trails are passable. Spring hiking can be difficult, high running creeks and swampy low areas can increase the difficulty of the hike. Mountain bikes, ATV's and motorcycles are not permitted on Park trails. --- Hiking and horses only.

See *Backpacking is Freedom* on page 13 for hiking tips.

Trails For the entire family!
Described in this guide are 30 one day trip or backpack trails. Distance of trails vary from 1.3 miles to 10 miles (One way). A easy short hike to historical Eugenia Mine or a difficult 8 mile climb above timberline to the keyhole at North Longs Peak --- trails for all skill levels.

Rocky Mountain National Park Index Map

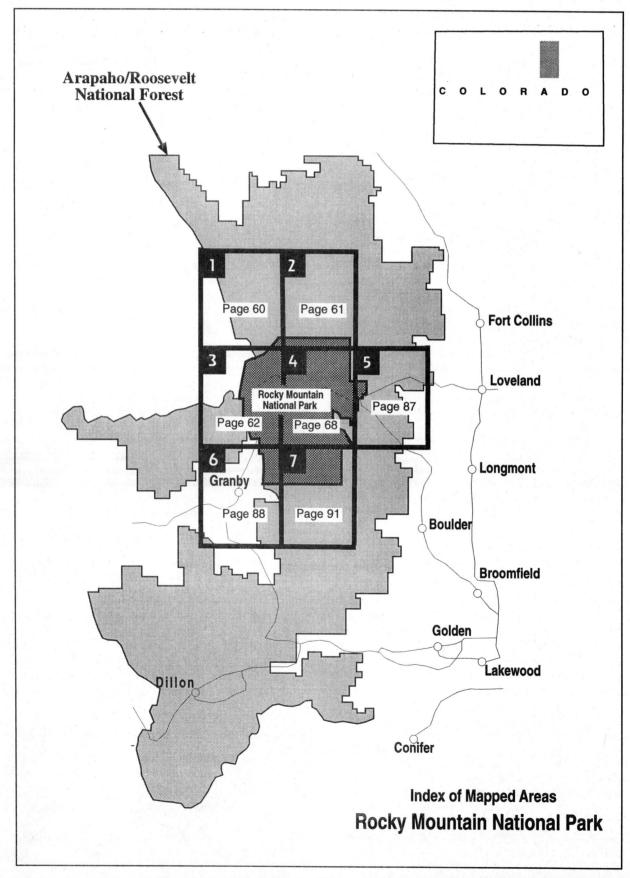

Arapaho/Roosevelt
National Forest

COLORADO

Fort Collins

Loveland

1 Page 60

2 Page 61

3

4 Rocky Mountain National Park
Page 62   Page 68

5 Page 87

6 Granby
Page 88

7 Page 91

Longmont

Boulder

Broomfield

Golden

Lakewood

Dillon

Conifer

Index of Mapped Areas
Rocky Mountain National Park

TABLE OF CONTENTS

Mechanized equipment is not permitted on trails within Rocky Mountain National Park. Vehicles must remain on roads or parking areas. Park only in designated parking areas. Camping is permitted only in designated areas. Pets are not allowed on trails or areas not accessible by automobile.

Text and photos of Rocky Mountain National Park Section of this guide
by
Don and Roberta Lowe.

Gorge on East Inlet Creek

No campgrounds located in the RMNP portion of this map

No trails described for RMNP portion of this map

Four Corners - 10 miles

McIntyre Creek
Drink Creek
966
4WD
963
Porter Creek
969
952
Rawah Creek
Lily Pond Lake
Roaring Creek
965
961
177
Lost Lake
Rawah Trail
Jimmy Creek
Lower Twin Lake
961
Green Ridge Trail
Williams Gulch
Boston Peak
McIntyre Lake
Springer Creek
984
968
103
2
968
972
Camp Cr.
968
Rapid Creek
Sleeping Elephant
Glen Echo & Rustic - 11 miles
978
Rawah Lakes
Sheep Mtn.
Upper Camp Lake
Fall Creek
West Branch
Half Mile Creek
Sleeping Elephant Mtn.
968
1
961
968
960
14
la Poudre
Bench Lake
North Fork
960
Cache
962
961
177
Rawah Wilderness
Laramie River
940
139
Twin Crater Lakes
West Fork
Twin Lakes
Comanche Peak Wilderness
Clear Creek
Clear Lake
960
Joe Wright Creek
3
Kelly Lake Trail
Carey Lake
959
Cameron Peak
103
177
Laramie Lake
4
Colorado
Island Lake
Timber Lake
Blue Lake
Lost Lake
May Creek
Kelly Lake
Hang Lake
Chambers Lake
Ties to Map 2   Page 61
Clark Peak
959
Fall Creek
5
Barnes Meadow Reservoir
Jewel Lake
State
Sawmill Creek
Grass Lake
944
939
Joe Wright Creek
Trap Creek
156
Peterson Lake
Forest
159
Joe Wright Reservoir
Bald Mtn.
Park
866
4WD
944
Montgomery Creek
977
Zimmerman Lake
Trap Creek
986
14
Trap Creek
Montgomery Pass
981
Neota Wilderness
Corral Creek
944
Walden - 27 miles
Diamond Peaks
6
Iron Mtn.
Long Draw Reservoir
R.M. N.P.
156

No campgrounds located in the RMNP portion of this map

No trails described for RMNP portion of this map

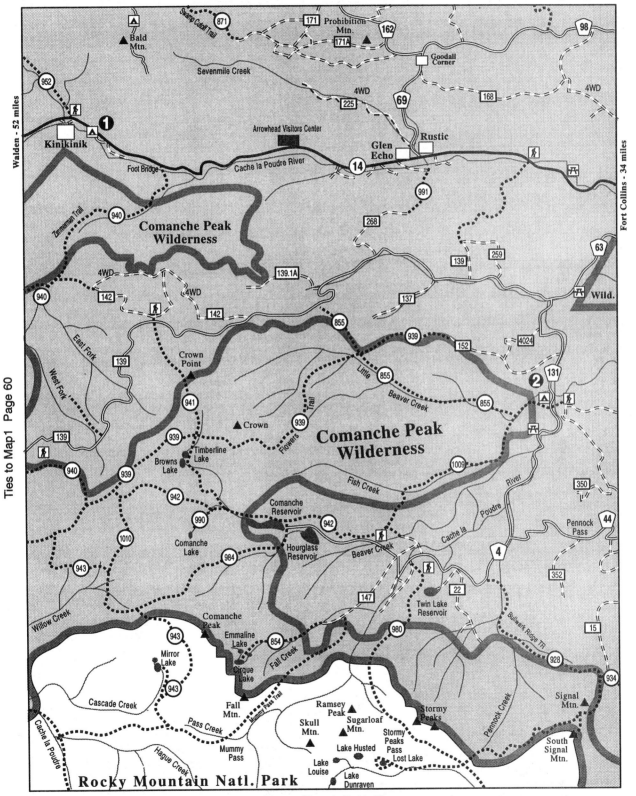

Ties to Map1 Page 60

Walden - 52 miles

Fort Collins - 34 miles

Ties to Map 4 Page 68

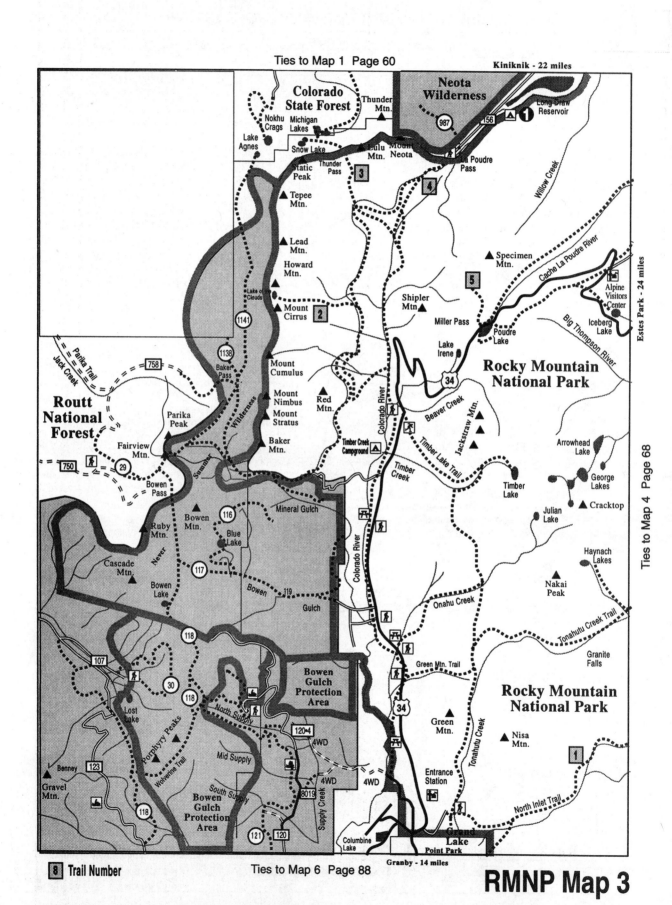

Ties to Map 1 Page 60

Kiniknik - 22 miles

Neota Wilderness

Colorado State Forest

Thunder Mtn.

Nokhu Crags

Michigan Lakes

Lake Agnes

Snow Lake

Static Peak

Thunder Pass

3

Lulu Mtn.

Mount Neota

La Poudre Pass

4

987

156

Long Draw Reservoir

Willow Creek

Tepee Mtn.

Lead Mtn.

Howard Mtn.

Lake of the Clouds

Mount Cirrus

2

Specimen Mtn.

Cache La Poudre River

5

Shipler Mtn.

Miller Pass

Poudre Lake

Lake Irene

Alpine Visitors Center

Iceberg Lake

Estes Park - 24 miles

Big Thompson River

1141

1138

758

Baker Pass

Mount Cumulus

Mount Nimbus

Mount Stratus

Red Mtn.

Baker Mtn.

Rocky Mountain National Park

34

Beaver Creek

Jackstraw Mtn.

Arrowhead Lake

George Lakes

Cracktop

Ties to Map 4 Page 68

Routt National Forest

Parika Trail

Jack Creek

Parika Peak

Fairview Mtn.

750

29

Bowen Pass

Wilderness

Summit

Timber Creek Campground

Timber Creek

Colorado River

Timber Lake Trail

Timber Lake

Julian Lake

Nakai Peak

Haynach Lakes

Ruby Mtn.

Never

Cascade Mtn.

116

Bowen Mtn.

Blue Lake

117

Bowen Lake

Bowen

119

Gulch

Mineral Gulch

118

Onahu Creek

Tonahutu Creek Trail

Granite Falls

107

30

118

Lost Lake

Porphyry Peaks

Wolverine Trail

North Supply

Mid Supply

Bowen Gulch Protection Area

Green Mtn. Trail

Green Mtn.

Rocky Mountain National Park

Nisa Mtn.

Tonahutu Creek

1

123

Benney

Gravel Mtn.

South Supply

Bowen Gulch Protection Area

121

120

8019

Supply Creek

120

4WD

4WD

4WD

Entrance Station

North Inlet Trail

Columbine Lake

Point Park

Grand Lake

Granby - 14 miles

8 Trail Number

Ties to Map 6 Page 88

Trail Number 1	Trail Name	Map Loc.	Distance	Difficulty	Beginning Elev.	Ending Elev	District
	North Inlet	E 6	10 mi	Difficult	8,472'	9,095'	RMNP

LAKES NOKONI AND NANITA.

ELEVATION GAIN: 2,580 feet, loss 300 feet. **HIGH POINT:** 11,080 feet. Allow 5 1/2 to 6 1/2 hours one way
ACCESS: Drive on U.S. 40 to the west end of Granby and turn north on U.S. 34. Proceed 14 miles to the junction with Colorado 278: keep right, following the sign to Grand Lake and Village. After one-third mile the road forks. Keep left, as indicated by the sign pointing to the Big Thompson Irrigation Tunnel and continue three-quarters mile to a marker on your left pointing to the Tonahutu Creek and North Inlet Trails. You can park along the shoulder or turn left, drive up the road, keeping left where it forks after 75 yards, and continue 200 yards to a signed parking area on the left. **ATTRACTIONS:** Many places in Colorado are named for Indians or use Indian words that describe some characteristic of the feature. The two lakes visited at the end of this scenic hike through woods and meadows are an example. Nokoni was one of the leading chiefs of the Comanche tribe and Nanita was either a Navajo word meaning plains Indians or the word used by a tribe of Texas Indians for the Comanches. These two lakes, situated in cirques near timberline, are separated by a 300 foot high ridge. Although long, the trail never climbs steeply and many sections are almost level. **NARRATIVE:** Walk along the road 200 feet beyond the parking area to a sign on your right stating North Inlet Trail. Drop for 75 feet and cross a large bridge a the boundary of Rocky Mountain National Park. Descend along the road bed through woods then level off and walk along the edge of a large meadow. Reenter woods then a 1.3 miles pass through an open area with picnic tables and outbuildings. Continue along the road for 150 yards to the guard station. A large sign just beyond the buildings lists several mileages and identifies the beginning of the trail proper.

The trail travels through woods at a gradual grade and at one point passes near a meadow. Near 3.0 miles begin climbing more noticeably and after traversing up a rocky slope come to a sign marking the path to Cascade Falls. Keep on the main trail and several hundred yards further pass a large campsite on your right. Climb moderately for a short distance then walk near the edge of a large meadow. Cross a small rocky slope that houses a colony of conies (rabbit) then resume walking through woods. Near 4.1 miles climb one switchback then walk briefly on the level before descending for a short distance. Wind above the gorges formed by turbulent North Inlet Creek and begin heading in an easterly direction. Hike at a moderate grade through woods and at 6.1 miles pass a large pond and meadow on your right.

Come to a small stream and several yards further cross Ptarmigan Creek on a foot log. Begin climbing more noticeably and make one short set of switchbacks. After a moderate uphill stretch make a second set and continue for one-half mile to the junction of the trail to Flattop Mountain. Further along the hike you will be able to see the long, level summit of this mountain and also the pyramid shape of Hallett Peak.

Keep right, drop slightly, then cross North Inlet Creek on a large bridge. Turn left and follow beside the stream for a short distance before curving right. Traverse along a wooded valley wall for one-third mile then begin a series of long switchbacks, coming at one point to an overlook at the edge of the deep gorge below Lake Nokomi. Climb along the steep, rocky wall of the canyon then curve right and walk through woods for a short distance to a flat area just before the shore where a sign points right to camping areas. To reach Lake Nanita turn left, cross the outlet creek and climb up then down the rocky ridge to the lake. **USGS:** Grand Lake, McHenrys Peak Quads. **RMNP MAPS:** 3 & 4.

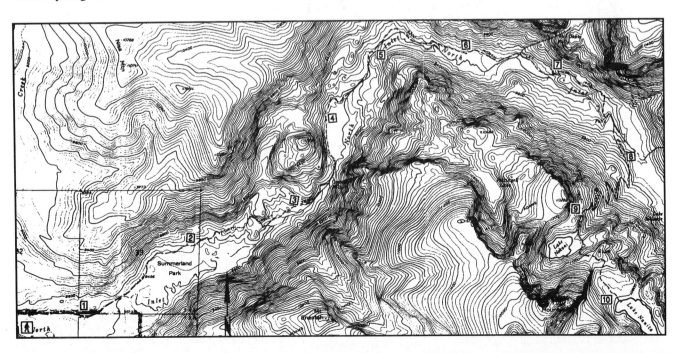

Trail Number 2	Trail Name	Map Loc.	Distance	Difficulty	Beginning Elev.	Ending Elev.	Ranger District
	Lake of the Clouds	E 5	6.3 mi	Difficult	9,095'	11,430'	RMNP

ELEVATION GAIN: 2,430 feet. **HIGH POINT:** 11,430 feet. Allow 4 to 4 1/2 hours one way.

ACCESS: Proceed on U.S. 34 9.5 miles north of the Grand Lake Entrance to Rocky Mountain National Park or 10.5 miles southwest from the Visitor Center at Fall River Pass to a sign stating Colorado River Trailhead Turn west and enter a parking area. At the trailhead hike one half mile to sign for Red Mountain Trail to left. **ATTRACTIONS:** Lake of the Clouds derives its name from the four peaks along the crest of the Never Summer Range, which are named after four different types of cloud formations -- Mounts Cirrus, Cumulus, Nimbus and Stratus. However, the lake rests in a basin well above timberline on the northeastern slope of Howard Mountain. Luke Howard, an early 19th Century English meteorologist, was the first to classify cloud formations and his name would have been a whimsical erudite choice for the peak. Most likely the mountain was named for a miner who worked in the area during the early 1880's. The final 0.5 mile of the hike is over a massive boulder field and an easily negotiated but steep slope. Although much more demanding than hiking along a trail, no special climbing skills are needed for this section. **NARRATIVE:** Cross the North Fork of the Colorado River on a foot bridge and walk 200 feet through a clearing to the western wall of the valley then begin climbing gradually through woods. After three-quarters mile cross an open, rocky slope where you can look down on a portion of the Kawuneeche Valley. (This name is a corrupted spelling of the Arapaho word for coyote.) Reenter woods and at 1.2 miles curve north and wind up at a moderate, but uneven, grade occasionally passing through rocky or small open areas. Begin a very gradual descent and come to an overlook where you will be able to see down to the parking area. Continue downhill to the crossing of Opposition Creek then resume climbing. The grade increases considerably just before reaching the Grand Ditch.Turn right and walk along the road that parallels the canal. The digging of this 14 mile agricultural irrigation channel was started in the late 1880's and the system still is maintained. Just beyond Lost Creek at 3.5 miles you will be able to look down onto the large meadow near the head of the Kawuneeche Valley, the side of Lulu City (See Thunder Pass Trail). About two miles beyond where you first met the road, be watching for a large bridge across the canal at Dutch Creek.

Climb through woods and soon pass the remains of two cabins. The grade becomes more moderate and the route continues winding up beneath conifers. Pass through a small meadow with a stream flowing through it and a short distance further walk along the shore of a tarn situated at the south end of a larger meadow. A campsite is located in the woods near the west edge of the clearing.

Resume the winding ascent and soon begin climbing more steeply. At the edge of the timber turn right and traverse up a low rocky bluff toward a large cairn. At the crest, where the trail ends, turn left (southwest) toward a 400 foot high rock wall with a waterfall running down its face. This flow is the outlet from Lake of the Clouds. Descend over the boulder field, generally aiming for the grassy shelves on the wall to the right of the falls. Where you resume climbing head toward the left hand edge of the benches, just above where the outlet begins to flow underground. Turn right and keep on the lower ledges until the slope ahead begins to fall away. Turn left and resume climbing steeply toward the crest. At the edge of the basin walk several hundred feet over grass and slabs to the shore of the lake. **USGS:** Fall River Pass, Mount Richthofen Quads. **RMNP MAP:** 3.

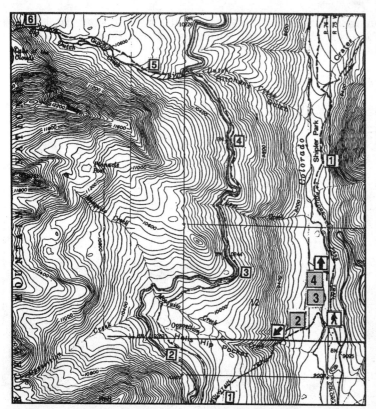

Boulder Field Below The Lake

	Trail Name	Map Loc.	Distance	Difficulty	Beginning Elev.	Ending Elev.	Ranger District
Trail Number **3**	**Thunder Pass**	E 5	6.5 mi	Difficult	9,095'	11,331'	RMNP

ELEVATION GAIN: 2,580 feet, loss 250 feet. **HIGH POINT:** 11,331 feet. Allow 3 1/2 to 4 1/2 hours one way.

ACCESS: Drive on U.S. 34 9.5 miles north of the Grand Lake entrance to Rocky Mountain National Park or 10.5 miles southwest from the Visitor Center at Fall River Pass to a sign stating Colorado River Trailhead. Turn west and enter a large parking area. A sign at the north edge of the turnaround stating Colorado River Trail and listing several mileages identifies the beginning of the hike. **ATTRACTIONS:** Lulu City, passed midway along the climb to Thunder Pass, was founded in 1879 and by 1882 had a population of 500 miners, shopkeepers and their families. The following year it was a ghost town, a victim of falling gold prices. Today, the remnants of a few log cabins mark the site. Another reminder of earlier times is the faint old wagon road the trail follows for portions of the hike. This road went north through the Kawuneeche Valley, over Thunder Pass, down past Michigan Lakes to Walden. The entire trail is pleasing visually as well as interesting historically. Thunder Pass is a large, rolling open crest and affords good views to the north and west. A scenic loop trip that would add only a few feet of elevation on gain and 2.5 miles can be made by walking along the Grand Ditch for 2.5 miles then descending along the rim of Little Yellowstone Canyon on the trail from La Poudre Pass.

The only campsites are at the Michigan Lakes outside the Park boundary.

NARRATIVE: Walk along the grass, brush and tree-covered floor of the valley, traveling beside the North Fork of the Colorado River for a short distance, then traverse the base of the valley wall. The grade is a series of level stretches interspersed with short sections of moderate ups and downs. Near 2.0 mile pass the remains of the two Shipler cabins, enter deeper woods and travel along a faint road bed. Curve gradually northeast and climb at a more steady grade. Come to the junction of the trail to La Poudre Pass. (If you make the recommended loop, you will be returning along this trail.)

Turn left and begin descending moderately, eventually switchbacking a few times. The trail becomes level and passes through a large meadow, the site of Lulu City. A short distance from the north end of the clearing keep left (straight) where the trail forks.

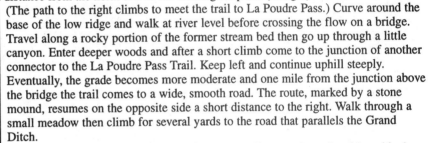

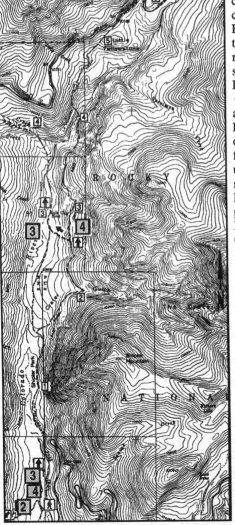

(The path to the right climbs to meet the trail to La Poudre Pass.) Curve around the base of the low ridge and walk at river level before crossing the flow on a bridge. Travel along a rocky portion of the former stream bed then go up through a little canyon. Enter deeper woods and after a short climb come to the junction of another connector to the La Poudre Pass Trail. Keep left and continue uphill steeply. Eventually, the grade becomes more moderate and one mile from the junction above the bridge the trail comes to a wide, smooth road. The route, marked by a stone mound, resumes on the opposite side a short distance to the right. Walk through a small meadow then climb for several yards to the road that parallels the Grand Ditch.

Cross the canal on a bridge and walk for a short distance along the old road bed at an almost level grade before climbing more noticeably. One-half mile from the Ditch go through the center of a large meadow. Reenter woods at the opposite side of the clearing and climb more steeply. After another 0.5 mile abruptly leave the forest and begin traversing up an open slope. Switchback to the right and continue up along the faint roadbed. Enter the small valley below the pass, wind through the swale at a very moderate grade then make the final, short climb to the crest at the Park boundary. The dirt road winds down to the Michigan Lakes just below the pass. The jagged crest of the Nokhu Crags pokes the sky to the northwest and the Medicine Bow Range fills much of the scene to the north. **USGS:** Fall River Pass Quad. **RMNP MAP:** 3.

Static Peak

Trail Number 4	Trail Name	Map Loc.	Distance	Difficulty	Beginning Elev.	Ending Elev.	Ranger District
	La Poudre Pass	E 5	7.0 mi	Difficult	9,095'	10,178'	RMNP

ELEVATION GAIN: 1,285 feet, loss 100 feet. **HIGH POINT:** 10,186 feet. Allow 4 to 4 1/2 hours one way.
ACCESS: Proceed on U.S. 34 9.5 miles north of the Grand Lake Entrance to Rocky Mountain National Park or 10.5 miles southwest from the Visitor Center Fall River Pass to a sign stating Colorado River Trailhead. Turn west and enter a large parking area. A sign at the north edge of the turnaround stating Colorado River Trail and listing several mileages identifies the beginning of the hike. **ATTRACTIONS:** Unlike most passes, La Poudre Pass is a long, flat bottomed, meadow filled valley. In fact, its earlier name was Mountain Meadows Pass. Scenically, the most impressive part of the hike is the 1.5 mile section above the fantastic rock formation of Little Yellowstone Canyon. Man-made remains can be seen by making a short side trip to Lulu City, for three years a busy mining town that once supported 500 people. A loop trip that would add 2.5 miles and less than 200 feet of elevation gain is possible by returning along the Thunder Pass Trail **NARRATIVE:** Walk along the grass, brush and tree covered floor of the valley, traveling beside the North Fork of the Colorado River for a short distance, then traverse the base of the valley wall. The grade is a series of level stretches interspersed with short sections of moderate ups and downs. Near 2.0 mile pass the remains of the two Shipler cabins. Joe Shipler was the first miner to settle in the valley. The main cabin, built in 1876, was occupied until 1914 and its sod roof did not collapse until 1963. Enter deeper woods and travel along a faint road bed. Curve gradually northeast and climb at a more steady grade. Come to the junction of the trail that goes down past Lulu City and climbs to Thunder Pass. (If you make the recommended loop, you will return along this trail.) Keep right and soon begin a very gradual descent. Weave in and out of little side canyons and during one stretch and large meadow that was the site of the once-thriving Lulu City can be seen below. Pass an unmarked downhill path on your left that meets the Thunder Pass Trail just north of Lulu City. Keep right, drop gradually to river level and walk near the flow before crossing it on foot bridges. Turn right and pass a sign identifying the beginning of Little Yellowstone Canyon. For the next several hundred yards the trail travels through extremely dense timber. Pass the junction of a second connecting path to the Thunder Pass Trail, keep right and cross a small stream on log foot bridges.

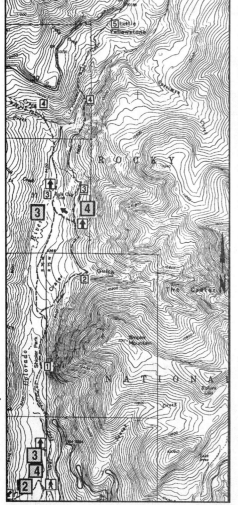

Climb at a moderate, steady grade, eventually coming close to the rim of the canyon. Cross the severely eroded ravine formed by Lady Creek on a bridge supported by a tall stone pillar and continue up along the wooded slope. Go in and out of a canyon and traverse a barren slope for a short distance. The grade lessens and the trail enters a second, larger canyon. Cross a creek at its head and drop slightly as you walk along the south facing slope. Travel almost on the level around the face of a ridge to the rim of the third and largest canyon. Leave the trail and walk a few feet toward the edge for a good view. Continue uphill at a steady grade then cross a small stream and begin a series of switchbacks. Wind up through an area of small clearings before climbing steeply for a short distance to the road that parallels the Grand Ditch. Turn right and follow the Ditch for 1.6 miles to the south end of La Poudre Pass where a bridge crosses the canal. A ranger station and a path to the campsites are on the opposite side of the span. To make the recommended loop, return to the point where you first met the road and continue in a southwesterly direction beside the Ditch for 2.5 miles. A few hundred feet beyond a cluster of buildings come to a foot bridge across the canal. Look for an unmarked trail across the road and follow it. (Refer to Thunder Pass Trail for detailed information for the remainder of the hike.) **USGS:** Fall River Pass Quad. **RMNP MAP:** 3.

Little Yellowstone Canyon

Trail Number 5	Trail Name	Map Loc.	Distance	Difficulty	Beginning Elev.	Ending Elev.	Ranger District
	Crater	E 5	1.5 mi	Easy	10,758'	11,540'	RMNP

ELEVATION GAIN: 782 feet. **HIGH POINT:** 11,540' feet. Allow 2 to 2 1/2 hours one way.
ACCESS: Drive on Trail Ridge Road (U. S. 34) 16 miles north of the Grand Lake Entrance of Rocky Mountain National Park or four miles southwest from the Visitor Center at Fall River Pass to a parking area on the north side of the highway a short distance southwest from the outlet end of Poudre Lake. **ATTRACTIONS:** The saddle at 1.5 miles is one of the best vantage points in the Park for spotting bighorn sheep. Even if you are not fortunate enough to see any of these agile animals, you will be surrounded by scenic terrain along the hike. Two thirds of the climb is above timberline and during the ascent have a good views of the Never Summer Range. Looking northwest from the narrow summit you can study the jagged spires of the Nokhu Crags (also called the Seven Utes) and beyond them to a portion of North Park. Closer to the southeast is the Visitor Center and the eastern portion of Trail Ridge Road above timberline and Longs Peak caps the skyline in the distance. Turning to the southwest, you may be able to see as far as the Gore Range. The trail climbs very steeply for much of its distance and no water is available at the trailhead or along the hike. **NARRATIVE:** The trail rises from the northwest side of the turnout through a grassy, open slope for several hundred feet to the register. Curve right and begin climbing very steeply through woods. Periodically, the trail grade moderates, but usually the route rises abruptly. Leave the timber at 1.1 miles and being a more gradual traverse along the tundra. Quarter of a mile beyond timberline come to a saddle at the base of the first peak where the trail ends. Interesting rock formations frame the view of the Never Summer Range to the west. The composition of the rocks here is evidence of the volcanic origin of Specimen Mountain. Bighorn sheep frequently graze on the western side of the summit ridge and you may be able to spot some by studying carefully the slopes below the saddle. **USGS:** Fall River Pass Quad. **RMNP MAP:** 3. **NOTE:** The trail formerly continued along the ridge to Specimen Mountain, but now ends as described in text.
THIS TRAIL IS CLOSED DURING BIGHORN SHEEP LAMBING SEASON. CALL BEFORE PLANNING TO USE THIS TRAIL.

Rain Over Never Summer Range

Ties to Map 2 Page 61

Rocky

Mountain

National

Park

Hazeline Lake

Flatiron Mtn.

Rowe Mtn.

Rowe Peak

Rowe Glacier

Hagues Peak

Mount Dunraven

Mount Dickenson

Desolation Peaks

Crystal Lake

Mummy Mtn.

Fairchild Mtn.

Lawn Lake

Ypsilon Mtn.

Spectacular Lakes

Ypsilon Lake

Chiquita Lake

Mount Chiquita

Chiquita Creek

Mount Chapin

Mount Tileston

Bighorn Mtn.

Dark Mtn.

West Creek

Cow Creek

MacGregor Mtn.

The Needles

Gem Lake

Marmot Point

Chapin Pass

Iceberg Pass

Fall River

Roaring River

Lawn Lake Trail

Horseshoe Falls

Horseshoe River

Aspenglen Campground

Entrance Station

Castle Mtn.

The Twin Owls

Estes Park

Cache La Poudre River

Chapin Creek

Tundra Trail

Sundance Mtn.

Ute Trail

Deer Ridge Junction

Deer Mtn.

Entrance Station

Forest Lake

Terrah Tomah Mtn.

Timberline Pass

Beaver Mtn.

Hayden Creek

Big Thompson River

Hidden River

Windy Gulch

Moraine Park Campground

Eagle Cliff Mtn.

Prospect Mtn.

Giant Track Mtn.

Marys Lake

Hayden Lake

Stones Peak

Rainbow Lakes

Spruce Lake

Fern Falls

Thompson

Cub Lake

Sheep Mtn.

Rams Horn Mtn.

Gabletop Mtn.

Fern Lake

Mount Wuh

Mill Creek

Bierstadt Lake

Glacier Basin Campground

Knobtop Mtn.

Lake Helene

Joe Mills Mtn.

Notchtop Mtn.

Flattop Mtn.

Bear Lake

Murphy Lake

Ptarmigan Pass

Emerald Lake

Lily Mtn.

Snowdrift Peak

Hallet Peak

Lake Haiyaha

Estes Cone

Otis Peak

The Loch

Mills Lake

Half Mtn.

Storm Pass

Twin Sisters Peaks

Water Dance Falls

Bench Lake

Andrews Glacier

Battle Mtn.

Twin Sisters Mtn.

North

Inlet Trail

Sky Pond

Glacier Gorge

Longs Peak Campground

Alpine Brook

Lake Nokoni

Taylor Peak

Taylor Glacier

Black Lake

Storm Peak

Chasm Lake

Inlet Creek

Lake Nanita

Lake Powell

Longs Peak

Tonahutu Creek Trail

Granby - 38 miles

Ties to Map 3 Page 62

Loveland - 30 miles

Lyons - 19 miles

Ties to Map 5 Page 87

Allenspark - 4 miles

No. Fork Trail

Grouse Creek

4WD

Twin Sisters Trail

Ties to Map 7 Page 91

2 Trail Number

Trail Number **6**	Trail Name	Map Loc.	Distance	Difficulty	Beginning Elev.	Ending Elev.	District
	Ypsilon Lake	F 5	5.0 mi	Mod/Diff	8,508'	10,600'	RMNP

ELEVATION GAIN: 2,085 feet, loss 200 feet. **HIGH POINT:** 10,710 feet. Allow 3 to 3 1/2 hours one way.
ACCESS: Proceed on Trail Ridge Road (U.S. 34)1.9 miles west of the Fall River Entrance to Rocky Mountain National Park or 1.5 miles north of Deer Ridge Junction to the sign identifying Fall River Road and parking for the Lawn Lake Trailhead. (Deer Ridge Junction is 2.8 miles west of the Beaver Meadows Entrance.) Turn north and after several hundred feet turn right and continue to the parking area where a small sign gives the mileages to Ypsilon and Lawn Lakes. **ATTRACTIONS:** Ypsilon is Greek for the letter Y and when viewed from a distance the long, snow filled erosion channels centered on the southeast face of Ypsilon Mountain unmistakably have this outline. Although sharing the name of the peak, Ypsilon Lake actually perches on a bench several hundred feet below timberline. For a closer view of the mountain amid more scenic terrain you can make the strenuous cross-country side trip to the Spectacle Lakes. This climb involves an additional one-half mile and 800 feet of elevation gain. **NARRATIVE:** The trail begins at a large bulletin board and although the path rises steeply for the first few hundred feet, the grade soon moderates. Where two trails join, first on the right then the left, continue climbing on the main route. After two sets of switchbacks resume traversing northwest along the wooded slope above Horseshoe Park. Near 0.5 mile begin curving into the side canyon formed by the Roaring River. Hike at a gradual grade then drop slightly to the junction of the trail to Lawn and Crystal Lakes

Turn left and descend to the bridge across the Roaring River. Walk on the level in a northerly direction for a few hundred yards then switchback left and traverse up the slope at a steep grade. Cross a level area where you will be able to look down on the Sheep Lakes and Horseshoe Park. Resume climbing at a steeper grade and hike for one mile up a narrow, wooded ridge crest. The grade becomes more moderate and the trail winds through less dense woods for 1.5 miles before traversing downhill for a few dozen yards to little Chipmunk Lake. Wind your way around boulders and trees and after a very short up and down across an open, rocky swath climb slightly before the final, steep descent to Ypsilon Lake. Cross the inlet creek to reach the numbered campsites.

If you plan to make the very steep climb to the Spectacle Lakes, follow the inlet creek to Ypsilon Lake up to the outlet stream from the lower Spectacle Lake. Cross the flows and climb steeply along the west side of the outlet for about one-quarter mile to the first lake. **USGS:** Trail Ridge Quad. **RMNP MAP:** 4.

Falls at Ypsilon Lake

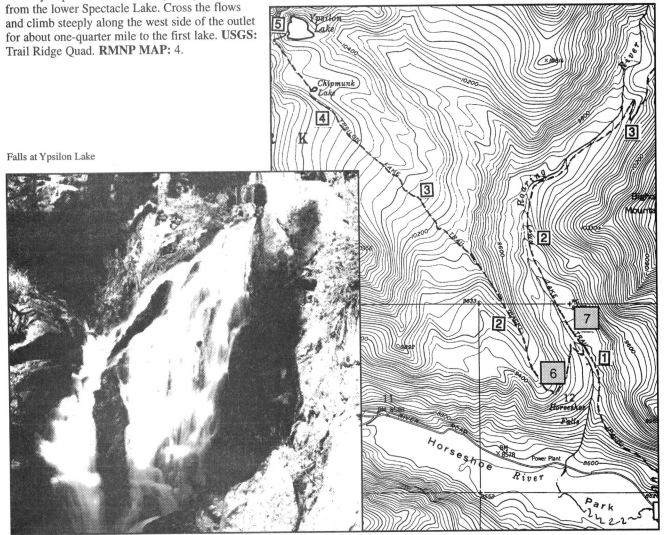

Trail Number 7	Trail Name	Map Loc.	Distance	Difficulty	Beginning Elev.	Ending Elev.	District
	Lawn & Crystal Lakes	F 5	7.5 mi	Mod/Diff	8,508'	10,987'	RMNP

ELEVATION GAIN: 2,900 feet. **HIGH POINT:** 11,550 feet. Allow 4 1/2 to 5 hours one way.
ACCESS: Drive on Trail Ridge Road (U. S. 34) one mile west of the Fall River Entrance to Rocky Mountain National Park or 1.5 miles north of Deer Ridge Junction to the sign identifying Fall River Road and parking for the Lawn Lake Trailhead. (Deer Ridge Junction is 2.8 miles west of the Beaver Meadows Entrance.) Turn north and after several hundred feet turn right and continue to the parking area where a small sign gives the mileage's to Ypsilon and Lawn Lakes. **ATTRACTIONS:** Crystal Lake, situated in the north central portion of Rocky Mountain National Park, is the headwaters of the Roaring River and except for the first one-half mile, the entire hike follows beside or near this turbulent flow. Although always scenic, the trail along the wooded slopes above the river is not as impressive as the final 1.5 miles between Lawn and Crystal Lakes. This latter section winds up over tundra to a basin filled with huge boulders and a cluster of tarns. **NARRATIVE:** The trail begins at a large bulletin board and although the path rises steeply for the first few hundred feet, the grade moderates. Where two trails join, first on the right then the left, continue climbing on the main route. After two sets of switchbacks traverse to the northwest on the wooded slope above Horseshoe Park. Near 0.5 mile begin curving into the side canyon formed by the Roaring River. Hike at a gradual grade then drop slightly to the junction of the trail to Ypsilon Lake

Keep right and continue walking at a very moderate grade near the Roaring River through both pure and mixed stands of aspen and conifers. As the grade increases the direction of travel changes gradually to the northeast. Just after crossing a small stream near 3.0 miles curve right and begin a long switchback. During the next mile the trail makes four sets of switchbacks.

Traverse along the valley wall of rocks and stunted trees then travel almost on the level before hiking above a small, open basin. Mummy Mountain is the large cliff face directly ahead to the north. Curve left, enter woods and begin climbing to the end of a grassy swale then walk through it to the slope above Lawn Lake. Although the shore is treeless and grassy in places, the smooth, green surface of the lake actually suggested the name to some fishermen who were visiting the site in the early 1870's. The dam across the outlet was built in 1903.

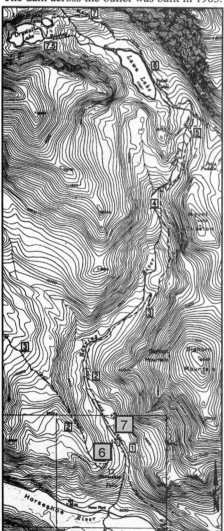

To reach Crystal Lake continue along the trail that climbs above the northeast side of Lawn Lake. Pass an outbuilding on your left near the end of the climb and descend to lake level. Resume climbing and wind up the open slope to the junction of the trail to Rowe Glacier at 7.1 miles. Keep left and after a short climb and drop resume a steady uphill grade. Cross a few small creeks and continue up through a little valley of boulders, slabs and grass where marmots are plentiful. Pass a large tarn nestled among the boulders then go by a second one before coming to Crystal Lake. **USGS:** Estes Park, Trail Ridge Quads. **RMNP MAP:** 4.

Lawn Lake

Trail Number **8**	Trail Name	Map Loc.	Distance	Difficulty	Beginning Elev.	Ending Elev.	District
	Gem Lake	G 5	2.0 mi	Moderate	7,720'	9,034'	RMNP

ELEVATION GAIN: 1,110 feet. **HIGH POINT:** 9,034 feet. Allow 1 1/2 hours one way.

ACCESS: Just east of the center of Estes Park turn north on MacGregor Avenue at the sign pointing to Devils Gulch and Glen Haven. Keep straight at the next intersection then after three-quarters mile curve right, staying on the main road. Three-quarters mile further be watching for a parking area off the road on your left (north). A sign stating Horse Trail - Gem Lake identifies the beginning of the hike. **ATTRACTIONS:** The climb to Gem Lake is the most easterly hike in Rocky Mountain National Park and the terrain and vegetation along the trip are much different than that only a few miles to the west on higher and less arid slopes. Particularly interesting are the smooth, rounded rock formations that comprise Lumpy Ridge on whose crest Gem Lake is tucked. Near the end of the hike you will have a panoramic view of the portion of the Continental Divide within the central region of the Park and an aerial like view of the city of Estes Park. Although the trip is short, the trail grade is moderately steep for much of its distance. Carry drinking water as the single source along the route may not be dependable. **NARRATIVE:** Walk on the level road through private property for 0.2 mile before beginning to travel on a wide trail. As you start to climb enter woods and pass near the first of the many weird rock formations you will enjoy along the route. Continue up an erratic and sometimes steep grade and at 0.7 mile enter Rocky Mountain National Park. A short distance from the boundary switchback up to your right. A path to the left at the turn leads to a small spring. Continue up and as you gain elevation you will have views of Lake Estes, Longs Peak and the area around Bear Lake.

Drop slightly then travel at a moderate uphill grade into a narrow canyon filled with aspen, conifers and varied rock groupings. Where the trail comes to a flat area and turns right, look left for a large, isolated rock with a hole through its upper end. The trail descends a few feet then resumes winding up through the canyon to the top of the ridge. Just before reaching the lake you will have a good view down onto Estes Park. In 1859 Joel Estes and one of his sons were the first white men to see then settle in the valley or "park" that today bears his name. Unlike most of the lakes in the Park, Gem Lake does not lie in a cirque but instead occupies a pocket on a ridge top surrounded by massive boulders. The lake has no inlet or outlet, the water level being maintained by seepage, rainfall and evaporation. **USGS:** Estes Park Quad. **RMNP MAP:** 4.

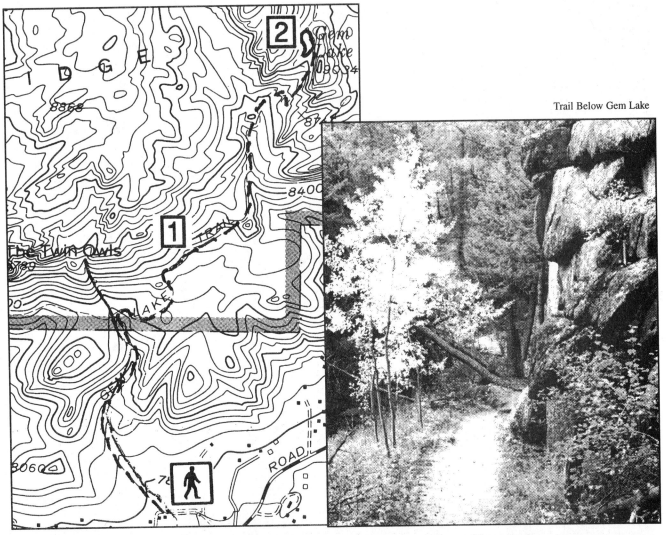

Trail Below Gem Lake

Trail Number 9	Trail Name	Map Loc.	Distance	Difficulty	Beginning Elev.	Ending Elev.	District
	Deer Mountain	G 5	1.6 mi	Moderate	8,900'	10,000'	RMNP

ELEVATION GAIN: 1,100 feet. **HIGH POINT:** 10,000 feet. Allow 1 1/2 hours one way.
ACCESS: The hike begins on the north side of the road about 100 yards east of the Deer Ridge Junction on Trail Ridge Road. This junction is 2.8 miles west of the Beaver Meadows Entrance to Rocky Mountain National Park and 2.4 miles west and south of the Fall River Entrance. Parking spaces are available along the shoulders of the highway. The actual trailhead is identified by a large sign stating Horse Trail - - Deer Mountain. **ATTRACTIONS:** Deer Mountain is the high point of the ridge that separates the Beaver Meadows and Fall River Entrances to Rocky Mountain National Park. Visitors to Horseshoe Park below the northwest end of Deer Ridge have an especially good view of the cluster of massive boulders comprising the summit of the mountain. However, the trail passes to the south of the rocky crest so you will need to climb cross-country for about 100 yards to reach the viewpoint.

Carry water as none is available along the hike. Although not especially high or exposed, the ridge frequently is struck by lightning during electrical storms.

NARRATIVE Drop slightly for 100 yards to the junction of the North Deer Mountain Trail to Estes Park. Keep right and climb gradually through the park like setting of widely-spaced trees. Descend for a short distance then begin a gradual uphill traverse along the grassy, aspen-dotted slope. Along this stretch you will have views down onto Moraine Park and across to Longs Peak and the area around Bear Lake. Switchback to the left and soon enter coniferous woods.

Begin a series of short, moderate, graded switchbacks. Eventually, you will be able to see down onto Horseshoe Park. Make a long traverse on the northwest slope of Deer Mountain and at 1.0 mile resume switchbacking. A short distance beyond the last switchback and where the trail begins traversing in an easterly direction at a gradual grade, turn left, leaving the established route. Climb cross-country for 100 yards to the rock outcroppings on the crest. From this perch you can enjoy views of the Mummy Range including Ypsilon Mountain to the northwest, Twin Sister Peaks and the Many Parks Curve area of Trail Ridge Road. The main trail continues at an almost level grade along the broad central portion of the crest then descends toward Estes Park. **USGS:** Estes Park Quad. **RMNP MAP:** 4.

Moraine Park at Longs Peak

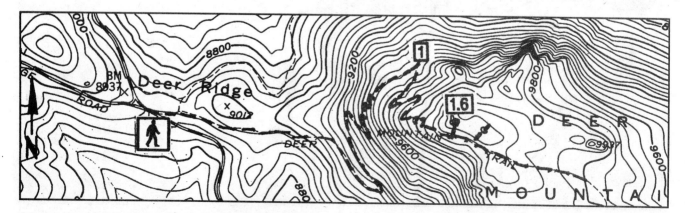

Trail Number 10	Trail Name	Map Loc.	Distance	Difficulty	Beginning Elev.	Ending Elev.	District
	Cub Lake & Mill Crk Basin	F 6	4.2 mi	Moderate	8,860'	8,600'	RMNP

ELEVATION GAIN: 1,270 feet, loss 430 feet. **HIGH POINT:** 9,420 feet. Allow 3 to 3 1/2 hours one way.
ACCESS: Proceed 0.1 mile west of the Beaver Meadows Entrance to Rocky Mountain National Park or 2.7 miles southeast of the Deer Ridge Junction of Trail Ridge Road (U.S. 34) to a sign pointing to Bear Lake and turn south. Drive downhill for one mile and turn right at the sign identifying the road to Moraine Park Campground and Fern Lake Trailhead. Proceed 0.25 mile and turn left 30 yards beyond a sign indicating the road to Cub and Fern Lakes Trailheads. The pavement ends after one mile near the parking area for a more easterly trail to Cub Lake and although level, the dirt road is very narrow. After another mile come to a parking area at the road's end. A sign at the west side of the turnaround marks the beginning of the Fern Lake Trail. **ATTRACTIONS:** Since spectacular alpine scenery is plentiful in Rocky Mountain National Park, the gentle setting of Mill Creek Basin is enjoyed as much for the pleasing contrast it affords as for its own beauty. The Basin is an aspen-splashed meadow surrounded by low wooded slopes and the landscape is especially attractive when the aspen bear the yellow, gold, and red leaves of fall. Mill Creek, and later the Basin, was named for the lumber mill that operated here in the late 1870's. A strenuous and scenic loop trip can be made by combining this hike with Fern, Odessa Lakes and Lake Helene Trails. The loop is 15 miles long and involves 3,000 feet of elevation gain. If you decide to make the circuit, the preferable route is to visit Fern and Odessa Lakes first, climb to Lake Helene then descend along Bierstadt Moraine to Mill Creek Basin. (Refer to Fern & Odessa Lake Trails.) **NARRATIVE:** Walk along the wooded valley floor, generally at a level grade but with a few minor ups and downs. Occasionally, the trail travels close to the Big Thompson River and at 1.5 miles the route passes under the immense boulders of Arch Rocks. One half mile further come to a large bridge at The Pool. Just beyond the west side of the span is the junction of the trail to Fern Lake. Keep left, go through woods for several yards to a side trail, turn left and pass a sign indicating the route to Cub Lake. Drop slightly for several feet then begin climbing moderately along the slope. The woods become quite dense along one stretch and the trail crosses two small streams. Briefly wind up through an open, rocky area then reenter woods and continue to the junction of the side trail to Cub Lake. To visit it, turn left and go through woods for one-third mile to a point above the shore. The west end of the lake usually is covered with lily pads. To continue the hike to Mill Creek Basin turn right at the junction and wind up at a steeper grade through coniferous woods sprinkled with aspen. At 3.0 miles begin traversing up the densely forested slope where you will have glimpses

down onto Cub Lake. As you near the top of the ridge the vegetation becomes less lush and at the crest there is no ground cover between the widely spaced pine trees. Descend, and after 0.5 mile begin traveling above the northwestern edge of the meadows of Mill Creek Basin. Come to the junction of a trail to Hallowell Park and keep right along an open slope. After several hundred feet pass an immense fireplace on your left and an outbuilding. The route continues around the edge of the meadow to the crossing of Mill Creek then comes to a junction where one trail goes east to Hallowell Park and the other heads southwest to Bierstadt and Bear Lakes. **USGS:** McHenrys Peak, Longs Peak Quads. **RMNP MAP:** 4.

Mill Creek Basin

Trail Number **11**	Trail Name	Map Loc.	Distance	Difficulty	Beginning Elev.	Ending Elev.	District
	Fern & Odessa Lakes	F 6	5.0 mi	Mod/Diff	8,860'	10,000'	RMNP

ELEVATION GAIN: 1,920 feet. **HIGH POINT:** 10,020 feet. Allow 3 to 3 1/2 hours one way.

ACCESS: Drive 0.1 mile west of Beaver Meadows Entrance to Rocky Mountain National Park or 2.7 miles southeast from the Deer Ridge Junction of Trail Ridge Road (U.S. 34) to a sign pointing to Bear Lake and turn south. Proceed downhill for one mile and turn right at the sign identifying the road to Moraine Park Campground. Travel one-quarter mile and turn left 30 yards beyond a sign identifying the road to Cub and Fern Lakes Trailheads. The pavement ends after one mile and although level, the dirt road is very narrow. After another mile come to a parking area at the road's send. A sign at the west side of the turnaround marks the beginning of the Fern Lake Trail. **ATTRACTIONS:** Like most hikes in the Park, the trail to Fern and Odessa Lakes travels through several different settings. For the first two miles the route parallels the Big Thompson River at the base of a steep rock wall then it climbs through deep woods to Fern Lake. One mile further you will reach Odessa Lake at the base of the Little Matterhorn. From the northeast it indeed does resemble a small version of the mountain that straddles the border between Italy and Switzerland.

If you have the time and energy you are urged to continue one mile beyond Odessa Lake along the steep wall of Joe Mills Mountain to Lake Helene. A strenuous loop trip down Bierstadt Moraine to Mill Creek Basin that would add eight miles and 1,200 feet of elevation gain also is possible.

NARRATIVE: Walk along the nearly level valley floor. At 1.2 miles the route passes under the immense boulders of Arch Rocks and one-half mile further comes to a large bridge at The Pool. Just beyond the west side of the span meet the junction of the trail to Cub Lake.

Keep right and several yards further keep right again. Climb a short distance then walk at a nearly level grade through woods. Pass a sign indicating campsites to the right and soon resume climbing. Cross Fern Creek on a bridge and wind loosely through woods then turn sharply left and traverse up a slope to Fern Falls. Switchback to the right and continue climbing through woods for 1.4 miles to the junction of the unimproved trail to Spruce Lake. Keep straight, climb 150 feet to the low ridge above Fern Lake and turn left.

Descend to the bridge across the outlet creek and walk along the shore. Travel through the edge of a large boulder field then reenter woods and begin climbing at a steady, moderate grade. At 4.9 miles come to a fork in the trail and keep right as indicated by the sign. Travel beside Fern Creek through a small canyon, cross a foot bridge and several yards further come to Odessa Lake.

To make the recommended side trip to Lake Helene, keep left at the fork at 4.9 miles and climb at a steady, moderate grade along the valley wall, soon leaving the timber. Near the crest at the head of the spectacular rocky gorge where the trail curves to the east be watching for a spur on your right to Lake Helene. To do the Mill Creek Basin loop trip, take the main trail east from Lake Helene as described in the text for Lake Helene and at the junction of the trail to Bear Lake at 0.4 mile follow the route to Bierstadt Lake. Soon begin descending and keep left at two junctions, following the signs pointing to Mill Creek. Turn left at the junction near the south edge of Mill Creek Basin and complete the circuit as described in the text for Cub Lake and Mill Creek Basin. **USGS:** McHenrys Peak Quad. **RMNP MAP:** 4.

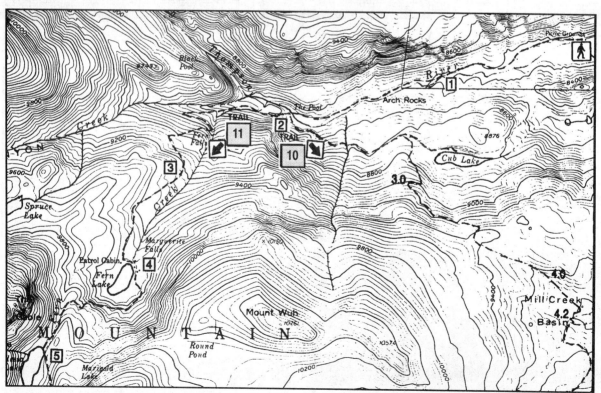

Trail Number 12	Trail Name	Map Loc.	Distance	Difficulty	Beginning Elev.	Ending Elev.	District
	Lake Helene	F 6	3.5 mi	Moderate	9,475'	10,690'	RMNP

ELEVATION GAIN: 1,440 feet. **HIGH POINT:** 10,690 feet. Allow 2 hours one way.
ACCESS: Proceed 0.1 mile west of the Beaver Meadows Entrance to Rocky Mountain National Park or 2.7 miles southeast from the Deer Ridge Junction of Trail Ridge Road (U.S. 34) to a sign pointing to Bear Lake and turn south. Drive downhill for one mile and keep straight where side roads go right to Moraine Park Campground and left to the museum. Continue eight miles to the end of the road at the Bear Lake Parking Area. Follow a sign at the east end of the turnaround indicating 100 feet to Bear Lake. **ATTRACTIONS:** Lake Helene lies in a picturesque basin directly below Flattop Mountain at the head of a spectacular rocky gorge. Small snow fields cling to the walls of the several cirques above the lake to the northwest between Flattop and Notchtop Mountains. You are urged to extend the hike by following the moderately graded trail that traverses down the precipitous eastern wall of the gorge for one mile to Odessa Lake at the base of Little Matterhorn (See Fern and Odessa Lakes). **NARRATIVE:** Where you come near the shore keep right and continue 50 feet along the paved path that circles the lake to a sign on your right stating Flattop Trail. Turn right and windup through an attractive aspen grove to the junction of the trail to Bierstadt Lake. Turn left and traverse up the slope that affords a view down onto Bear Lake and across to Longs Peak during an open stretch. Cross to the opposite side of the ridge and walk on the level above a cluster of boulders before climbing a short distance in woods to the junction of the trail to Flattop Mountain.

Keep right and climb moderately through woods then traverse above the head of Bierstadt Moraine that stretches out to the northeast. As you gain elevation you will be able to see Bierstadt Lake near the eastern end of the flat crest. At the edge of a grassy, open area come to a stream and an inviting snack stop. Continue up at an easy grade just at the edge of timber-line. Briefly hike in deeper woods before traversing a rocky slope then resume traveling through conifers rs to dry Mill Creek.

Wind up a slope of boulders, grass and trees then travel along a wooded slope, passing a sign identifying Two Rivers Lake below. Walk on the level for a short distance then drop very gradually for 0.1 mile to a sign marking the side path to Lake Helene, a portion of which you can see from the junction. Keep left and descend about 100 yards to the brushy north shore.

To make the recommended trip along the steep western wall of Joe Mills Mountain, named for Enos Mills brother, to Lake Odessa, keep right on the main trail and after a short distance curve north into the rocky gorge.

USGS: McHenrys Peak Quad. **RMNP MAP:** 4.

Notchtop Mountain

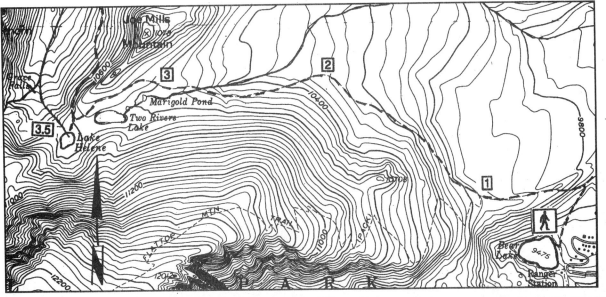

Trail Number 13	Trail Name	Map Loc.	Distance	Difficulty	Beginning Elev.	Ending Elev.	District
	Bierstadt Lake	F 6	1.3 mi	Easy	8,860'	9,440'	RMNP

ELEVATION GAIN: 600 feet. **HIGH POINT:** 9,440 feet. Allow 3/4 to 1 hour one way.
ACCESS: Drive 0.1 mile west of the Beaver Meadows Entrance to Rocky Mountain National Park or 2.7 miles southeast from the Deer Ridge Junction of Trail Ridge Road (U.S. 34) to a sign pointing to Bear Lake and turn south. Proceed downhill for one mile and keep straight where side roads go right to Moraine Park Campground and left to the museum. Continue five miles to a sign identifying the parking area for the Bierstadt Lake Trail on the right (north) side of the road. The trail begins at the northwest edge of the parking area. **ATTRACTIONS:** Bierstadt Lake was named for Albert Bierstadt, a noted American artist who painted in the area in the 1870's. The lake rests on the top of massive Bierstadt Moraine. Along the first half of this short trail, aspen trees frame excellent views of the rugged back range.

The glaciers that deposited the debris comprising Bierstadt and the other moraines in the Park and carved the cirques and U-shaped valleys were, geologically speaking, recent performers in the pageant of forces that created the land forms seen today. At the time when the simplest life forms were beginning on earth the great sea that covered this region of the present day Rockies was beginning to recede and mountains slowly were rising. Eventually, the first mountains were worn away to a rolling plain and near the end of the Age of Reptiles about 60 million years ago the Rocky Mountains of today began rising. One million years ago

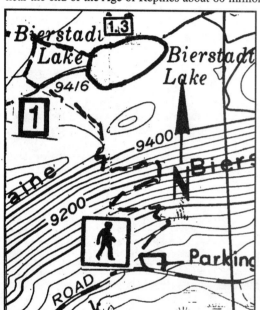

the Ice Age began and glaciers formed in the high mountains. (The glaciers in the Colorado Rockies were not part of the massive fields of ice that covered much of North America during this period.) These glaciers receded, leaving behind the cirques, lakes, tarns, moraines and U-shaped valleys seen today. Only a few small glacial remnants remain in the Park and Flattop Mountain and Andrews Tarn Trails visit two of them. **NARRATIVE:** Climb for a few yards and just beyond the infromation sign curve right and continue through deep evergreen woods at a moderate, steady grade. As the trail begins switchbacking the conifers are replaced by aspen. Climb through an open area that affords especially good views then resume switchbacking up through deciduous woods and occasional clearings.

Pine and spruce resume their dominance at the crest of the moraine and the trail descends very gradually for a short distance. The grade soon becomes level and during one stretch the trail is along a dirt dike. Just beyond this built-up section come to the junction of the trail to Bear Lake. Keep right, walk along another dike and come to a second junction. Turn right and descend slightly. At a sign pointing to Bear Lake keep straight (right) and a short distance further keep right again at an unsigned fork to the shore of Bierstadt Lake. **USGS:** Longs Peak, McHenrys Peak Quads. **RMNP MAP:** 4.

Hallet Peak From Bierstadt Lake Trail

Trail Number 14	Trail Name	Map Loc.	Distance	Difficulty	Beginning Elev.	Ending Elev.	District
	Flattop Mountain	F 6	4.5 mi	Difficult	9,475'	12,324'	RMNP

ELEVATION GAIN: 3,075 feet. **HIGH POINT:** 12,324 feet. Allow 3 to 3 1/2 hours one way.

ACCESS: Proceed 0.1 mile west of the Beaver Meadows Entrance to Rocky Mountain National Park or 2.7 miles southeast from the Deer Ridge Junction of Trail Ridge Road (U.S. 34) to a sign pointing to Bear Lake and turn south. Drive downhill for one mile and keep straight where side roads go right to Moraine Peak Campground and left to the museum. Continue eight miles to the end of the road at the Bear Lake Parking Area. Follow a sign at the east end of the turnaround stating Bear Lake 100 feet. **ATTRAC-TIONS:** The trail to the broad, level summit of Flattop Mountain climbs at a steady, moderate grade past a succession of scenic attractions. The final 1.5 miles are above timberline and during this portion you will have views to the east over Bierstadt Moraine and beyond to Estes Park and northwest to the highest sections of Trail Ridge Road. From the summit you can continue south a few hundred yards and have lunch at the edge of Tyndall Glacier, just below the crest. Hikers wanting a more strenuous trip can continue along the gentle western side of the Continental Divide and climb Hallett Peak or visit Andrews Glacier.

   NARRATIVE: Where you come near the shore of Bear Lake keep right and continue 50 feet along the paved path that circles the lake to a sign on your right identifying the Flattop Trail. Turn right and wind up through an attractive aspen grove to the junction of the trail to Bierstadt Lake. Turn left and traverse up the slope that affords a view down onto Bear Lake and across to Longs Peak during an open stretch. Cross to the opposite side of the ridge and walk on the level above a swale of boulders before climbing a short distance in woods to the junction of the trail to Lake Helene. Turn left and begin a series of switchbacks through coniferous woods to the viewpoint above Dream Lake (See Emerald Lake Trail). Mills Lake sprawls on a bench across the valleys to the south (refer to Black Lake Trail).

   Continue up through woods and at 2.7 miles come to timberline. About 100 yards from where the trees are replaced by brush and gnarled, stunted conifers. A short distance beyond begins a series of switchbacks and at 3.0 miles pass the viewpoint 1,200 feet above Emerald Lake. As you look to the south you can see that the rock mass often identified as Hallett Peak from Bear Lake is really the lower part of a ridge that, after rising another 700 feet, ends in its actual, and much less imposing summit.

   From the viewpoint continue climbing along the mountain side, where conies (rabbit) are plentiful. Curve to the less steep northeast slope and traverse at a gradual grade for about one mile. Pass a metal hitching post and continue the final few hundred yards to the flat summit area.

   To reach Hallett Peak continue along the almost level trail and keep left at an unmarked fork. Walk south around the Tyndall Glacier cirque and make the easy climb to the summit from the west. To reach the viewpoint above Andrews Glacier and Tarn, continue south along the crest past the head of Chaos Canyon and Otis Peak for one and three-quarter miles to Andrews Pass **USGS:** McHenrys Peak Quad. **RMNP MAP:** 4.

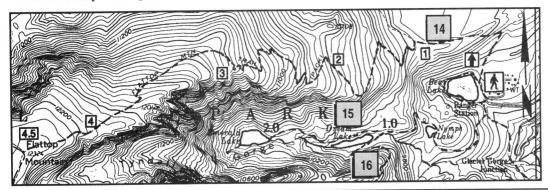

Longs Peak From Flattop Mtn Trail

Trail Number 15	Trail Name	Map Loc.	Distance	Difficulty	Beginning Elev.	Ending Elev.	Agency
	Emerald Lake	F 6	1.8 mi	Easy/Mod	9,475'	10,100'	RMNP

ELEVATION GAIN: 650 feet. **HIGH POINT:** 10,100 feet. Allow 1 to 1 1/2 hours one way.
ACCESS: Drive 0.1 mile west of the Beaver Meadows Entrance to Rocky Mountain National Park or 2.7 miles southeast from the Deer Ridge Junction of Trail Ridge Road (U.S. 34) to a sign pointing to Bear Lake and turn south. Proceed downhill for one mile and keep straight where side roads go right to Moraine Park Campground and left to the museum. Continue eight miles to the end of the road at the Bear Lake Parking Area. **ATTRACTIONS:** The trail to Emerald Lake passes little Nymph Lake, profusely covered with lily pads, and long, narrow Dream Lake as it penetrates Tyndall Gorge, situated between the eastern ridges of Hallett Peak and Flattop Mountain. **NARRATIVE:** Follow a sign at the east end of the turnaround stating Bear Lake 100 feet. Where you come near the shore turn left at the sign pointing to Emerald Lake Trail. Keep straight for a few hundred feet to the junction of the trail to Glacier Gorge Junction and keep right, following the arrows pointing to Nymph, Dream and Emerald Lakes. The trail is paved for the first one-half mile and climbs very moderately through woods. Come to Nymph Lake and walk around its east and north sides. Climb more noticeably, turn left and traverse 150 feet above the lake. Curve to the west along the face of an open slope and travel up among boulders and small aspens above a rugged little slope. Just before a stream at 1.0 mile take the trail up to the right and meet the junction at a bridge of the trail to Lake Haiyaha.

Keep right and after 0.1 mile come near the east end of Dream Lake. Walk parallel to the lake on the main trail and eventually travel close to the water's edge. At the west end of the lake follow the path to the right and after a few yards scramble for several feet up along a low rock band. Wind up and sometimes down, through deep woods. Cross a little stream and scramble up rocks for a few feet. As you near the exit creek from Emerald Lake veer right through a tight little thicket of evergreens. Keep right where the trail appears to curve left around a large boulder and walk away from the stream across a small open bench. Traverse up a low rock band and again travel on a well-defined trail then squeeze between two huge boulders at the east end of Emerald Lake. The

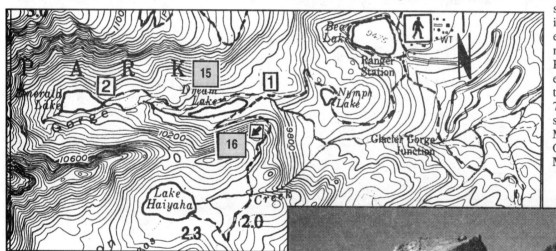

sheer rock face rising above the west end of the lake is not actually Hallett Peak but rather the east end of the ridge that continues up for 700 feet to the true summit. **USGS:** McHenrys Peak Quad. **RMNP MAP:** 4.

Hallett Peak from Bear Lake

Trail Number 16	Trail Name	Map Loc.	Distance	Difficulty	Beginning Elev.	Ending Elev.	Agency
	Lake Haiyaha	F 6	2.3 mi	Easy/Mod	9,475'	10,240'	RMNP

ELEVATION GAIN: 810 feet, loss 100 feet. **HIGH POINT**: 10,240 feet. Allow 1 to 1 1/2 hours one way.
ACCESS: Proceed 0.1 mile west of the Beaver Meadows Entrance to Rocky Mountain National Park or 2.7 miles southeast from the Deer Ridge Junction of Trail Ridge Road (U. S. 34) to a sign pointing to Bear Lake and turn south. Drive downhill for one mile and keep straight where side roads go right to Moraine Park Campground and left to the museum. Continue eight miles to the end of the road at the Bear Lake Parking Area. **ATTRACTIONS:** Haiyaha is an Indian word for rock and the name was appropriately applied to Lake Haiyaha as the shores are surrounded by massive boulders. The hike to the lake can be done as a scenic loop if you want a slightly more strenuous trip. This circuit would add 1.8 miles and gain 300 feet.
NARRATIVE: Follow a sign at the east end of the turnaround indicating 100 feet to Bear Lake. Where you come near the shore turn left at the sign pointing to Emerald Lake Trail. Keep straight for a few hundred feet to the junction of the trail to Glacier Gorge Junction. (If you make the recommended loop you will be returning along this trail.) Keep right, following the arrows pointing to Nymph, Dream and Emerald Lakes. The trail is paved for the first one-half mile and climbs very moderately through woods. Come to Nymph Lake and walk along its east and north sides. Climb more noticeably, turn left and traverse 150 feet above the lake. Curve to the west along the face of an open slope and hike above a rugged little slope among boulders and small aspens. At 1.0 mile, just before a stream, take the trail up to the right and come to the junction of the trail to Dream and Emerald Lakes at a bridge.

Keep left and cross the span, bear left where the trail forks. Curve right and begin traversing up the densely wooded slope above Dream Lake. After climbing for 0.2 mile switchback to the left and continue up through woods. Curve right around the steep face of the ridge. Since the trees are sparse along this stretch, you can see down onto several lakes, including Bear and Nymph, and across to Longs Peak. Hike along the southeast side of the ridge and soon begin a gradual descent.

Cross the outlet from Lake Haiyaha on a footbridge then ford a smaller stream. Travel on the level for a short distance to the junction of the trail to Loch Vale. Keep right and climb very moderately, passing a tarn on your left and crossing a short stretch of boulders before coming to the rocky shore of the lake.

To make the recommended loop, keep left at the junction to Loch Vale and begin descending. The trail eventually levels off and 1.5 miles from the fork come to the four-way junction of the trails to Mills Lake, Lock Vale and Glacier Gorge Junction. Take the far left trail and after a short climb and a traverse wind down to the large mileage sign just above the Glacier Gorge Junction Parking Area. (Refer to Sky Pond Trail for a detailed description of this section.) Keep straight following the arrow pointing to Bear Lake and climb for one-third mile to your starting point. **USGS:.** McHenrys Peak Quad. **RMNP MAP:** 4.

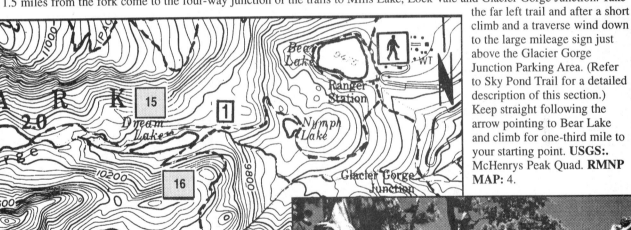

Gnarled Tree at Lake Haiyaha

Trail Number 17	Trail Name	Map Loc.	Distance	Difficulty	Beginning Elev.	Ending Elev.	Agency
	Andrews Tarn	F 6	4.6 mi	Difficult	9,475'	11,380'	RMNP

ELEVATION GAIN: 2,180 feet. **HIGH POINT:** 11,380 feet. Allow 3 to 3 1/2 hours one way.
ACCESS: Drive 0.1 mile west of the Beaver Meadows Entrance to Rocky Mountain National Park or 2.7 miles southeast from the Deer Ridge Junction of Trail Ridge Road (U.S. 34) to a sign pointing to Bear Lake and turn south. Proceed downhill for one mile and keep straight where side roads go right to Moraine Park Campground and left to the museum Continue 7.5 miles to the Glacier Gorge Parking Area. The trailhead is south across the road just above the steep curve of the highway and is marked by a large sign. **ATTRACTIONS:** Tucked into niches on the east side of the peaks in Rocky Mountain National Park are five small remnants of the much larger glaciers that once crept down the slopes. Collecting melt water from Andrews Glacier, the second most southerly of the five, is a small lake of the same name perched on the edge of an only slightly larger bench. It is reached by a path that travels along the rocky floor of a spire-rimmed valley then winds steeply up to the tarn.

A side trip to Glass Lake and a view over Loch Vale would add only one mile and 250 feet of elevation gain. (Refer to Sky pond Trail)

NARRATIVE: Turn left at the sign, cross a bridge and soon come to the junction of a spur to the Storm Pass Trail. Keep right and climb quite moderately, first crossing a small bridge then a larger one. Pass Alberta Falls at 0.5 mile and windup to the junction of the North Longs Peak Trail (No. 49). Keep right and continue climbing for a short distance then begin contouring above a canyon. Descend for several yards and come to the four way junction of the trails to Black Lake and Haiyaha and Bear Lakes. Take the middle of the three routes and traverse up at a moderate grade, soon entering the little gorge formed by Icy Brook. Climb in two complete sets of switchbacks then make a sweeping curve and come to the northeast end of The Loch. Turn right and enter a small open area where several side paths branch from the main route. Keep left at the first fork then, bear right and a few feet further

stay left. Travel around the north shore of The Loch in a series of small ups and downs then begin climbing through woods. At 3.5 miles come to the junction of the trail to Glass Lake and Sky Pond, turn right and hop across a small stream. Wind steeply up through woods for a few hundred yards then climb more moderately. Enter a valley and where a trail joins from your right near a meadow note the location of the junction so you will not miss it on the return. Hike along the southern slope of the valley and soon leave the deep timber. Traverse an area of small boulder fields, following circles of orange spray paint on the rocks, then just after crossing a small stream take the trail that climbs to the left. Follow the path along the floor of the rubble-strewn valley, keeping the stream on your left. At the base of the large rock mass in the center of the basin near the head of the valley cross the stream and wind up the gully to the south of the rocks. After climbing steeply for 0.2 mile come to Andrews Tarn at the edge of a small bench. **USGS:** McHenrys Peak Quad. **RMNP MAP:** 4.

The Sharkstooth

Trail Number 18	Trail Name	Map Loc.	Distance	Difficulty	Beginning Elev.	Ending Elev.	Agency
	Sky Pond	F 6	4.5 mi	Mod/Diff	9,475'	10,800'	RMNP

ELEVATION GAIN: 1,700 feet. **HIGH POINT:** 10,800 feet. Allow 2 1/2 to 3 hours one way
ACCESS: Proceed 0.1 mile west of the Beaver Meadows Entrance to Rocky Mountain National Park or 2.7 miles southeast from the Deer Ridge Junction of Trail Ridge Road (U.S. 34) to a sign pointing to Bear Lake and turn south. Drive downhill for one mile and keep straight where side roads go right to Moraine Park Campground and left to the museum. Continue 7.5 miles to the Glacier Gorge Parking Area. The trailhead is south across the read just above the steep curve of the highway and is marked by a large sign. **ATTRACTIONS:** Lock Vale, the valley through which the trail travels for the last two miles of the hike, was named for a Mr. Locke but the spelling was changed to Loch, the Scottish word for lake. The final one third mile from Glass Lake (also more pleasingly called Lake of Glass) to Sky Pond is along a level but very rocky cross-country route. However, for those who do not enjoy boulder hopping, Glass Lake affords a good stopping place with a fine view. Since the short, but steep, path to Andrews Tarn leaves the main route at 3.5 miles, you easily could combine the two hikes. This side trip would add a total of two miles and 1,000 feet of elevation gain. **NARRATIVE:** Turn left at the sign and cross a bridge. Pass a beaver pond on your left and at the register come to the junction of a spur to the Storm Pass Trail. Keep right and climb moderately, soon crossing a small bridge then a larger one. Continue rising beneath conifers and aspen and pass Alberta Falls at 0.5 mile. Wind up, sometimes traveling close to the narrow, deep rocky gorge formed by Glacier Creek, to the junction of the North Longs Peak Trail.

Keep right and continue climbing for a short distance then begin contouring around the most easterly of the two Glacier Knobs. Descend for several yards and come to the four-way junction of the trails to Black Lake and Haiyaha and Bear Lakes. Take the middle of the three routes and traverse up at a moderate grade, soon entering the little gorge formed by Icy Brook. Climb in two complete sets of switchbacks then make a sweeping curve and come to the northeast end of The Loch. Turn right and enter a small open area where several side paths branch from the main route. Keep left at the first fork then bear right and a few feet further stay left. Travel around the north shore of The Loch in a series of small ups and downs then begin climbing through woods. Pass under a huge overhanging rock face at the edge of Icy Brook and 0.1 mile further come to the junction of the trail to Andrews Tarn.

Keep left, drop a few feet and hop across a small stream. Hike through woods and small clearings to the base of a rocky steep slope. Wind up the steep slope then scramble up the rock gully beside Timberline Falls. During the climb you will be able to see back down to The Loch. Simultaneously, come to the edge of the basin, timberline and Glass Lake. To complete the hike travel near the northwest shore of Glass Lake then continue scrambling over small cliffs and large boulders to the barren north end of Sky Pond. **USGS:** McHenrys Peak Quad. **RMNP MAP:** 4.

Timberline Falls

Trail Number 19	Trail Name	Map Loc.	Distance	Difficulty	Beginning Elev.	Ending Elev.	Agency
	Black Lake	F 6	5.0 mi	Mod/Diff	9,475'	10,620'	RMNP

ELEVATION GAIN: 1,400 feet. **HIGH POINT:** 10,620 feet. Allow 3 1/2 to 4 hours one way.
ACCESS: Drive 0.1 mile west of the Beaver Meadows Entrance to Rocky Mountain National Park or 2.7 miles southeast from the Deer Ridge Junction of Trail Ridge Road (U.S. 34) to a sign pointing to Bear Lake and turn south. Proceed downhill for one mile and keep straight where side roads go right to Moraine Park Campground and left to the museum. Continue 7.5 miles to the Glacier Gorge Parking Area. The trailhead is south across the road just above the steep curve of the highway and is marked by a large sign. **ATTRACTIONS:** The east side of the Continental Divide between Flattop Mountain and Longs Peak was a region of heavy glaciation and the exceptionally scenic resuits are visited by a network of five trails. Black Lake is the most southerly destination in this group and lies beneath the almost vertical walls of McHenrys Peak at the head of three mile long Glacier Gorge.

Not difficult to negotiate, backpackers and hikers wanting a more strenuous trip can climb for 0.75 mile above Black Lake, gaining about 1,000 feet of elevation, to the superb setting of Frozen Lake.
NARRATIVE: Turn left at the sign, cross a bridge and soon come to a junction of the spur to the Storm Pass Trail. Keep right and climb quite moderately, soon crossing a small bridge then a larger one. Pass Alberta Falls at 0.5 mile and wind up to the junction of the North Longs Peak Trail.

Keep right and continue climbing for a short distance then travel on the level above a canyon. Wind down for several yards and come to the four-way junction of trails to Sky Pond and Andrews Tarn and to Haiyaha and Bear Lakes. Keep left on the most easterly trail and soon drop slightly to the bridge over Icy Brook. Walk at a moderate grade then climb along a rocky slope for several yards to an area of slabs. Turn left as indicated by the cairns and after a short level stretch, descend to cross Glacier Creek on a bridge. Turn right and climb for several hundred feet through a little aspen grove then walk over a level area of slabs and boulders to the end of Mills Lake.

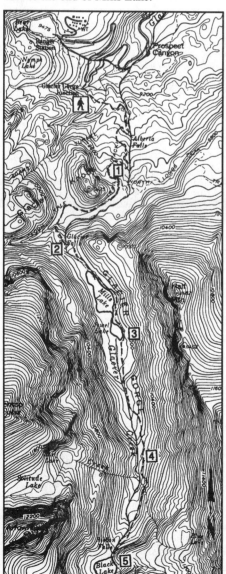

The trail continues near the east shore of Mills Lake and where the route forks midway along its length you can follow either one. At the end of the lake come to a sign identifying Jewel Lake to the west. Beyond the small lake cross a marshy meadow on a log walkway then wind up through woods at an erratic grade with a few slight drops. Come to a small patch of boulders at the edge of a meadow. Cross the rocks and walk beside a wide stream for several yards then resume climbing along the slope.

Pass below a little waterfall at 4.2 miles and continue traversing through woods and meadows. Walk on the level through an open area then wind up a rock band beside Ribbon Falls. At the crest cross a small grassy area then some flat stones and come to the edge of Black Lake. Scramble over boulders along the northeast shore for several yards then walk through woods above the water before dropping to a clearing near an inlet creek.

To make the cross country trip to Frozen Lake, climb beside the inlet creek near the campground and as the slope becomes more gradual curve southwest, pass under the face of the Spearhead and come to the lake's northeastern shore.
USGS: McHenrys Peak Quad. **RMNP MAP:** 4.

McHenrys Peak from Black Lake

Trail Number 20	Trail Name	Map Loc.	Distance	Difficulty	Beginning Elev.	Ending Elev.	Agency
	North Longs Peak	F 6	8.0 mi	More Diff	8,840'	13,500'	RMNP

ELEVATION GAIN: 4,400 feet. **HIGH POINT:** 13,500 feet. Allow 5 to 6 hours one way.
ACCESS: Proceed 0.1 mile west of the Beaver Meadows Entrance to Rocky Mountain National Park or 2.7 miles southeast from the Deer Ridge Junction of Trail Ridge Road (U.S. 34) to a sign pointing to Bear Lake and turn south. Drive downhill for one mile and keep straight where side roads go right to Moraine Park Campground and left to the museum. Continue 7.5 miles to the Glacier Gorge Parking Area. The trailhead is south across the road just above the steep curve of the highway and is marked by a large sign. **ATTRACTIONS:** Indians, and later trappers, used Longs Peak, the highest in northern Colorado and its neighbor Mt. Meeker as twin reference points when traveling on the plains. The natives called the peaks the Two Guides and the French trappers referred to them as The Two Ears. In 1820 Major Stephen Long led a party of 22 men along the Front Range. The map of his expedition showed Longs Peak as Highest Peak, but common usage soon affixed the Major's name to the mountain.

The established trail ends high above timberline at Boulder Field, a gently sloping expanse of huge rocks on the north shoulder of Longs Peak. Two short, but strenuous, side trips continue over boulders to spectacular viewpoints at the Keyhole and Chasm View, 1,700 feet directly above Chasm Lake. The weather can be especially inclement on Longs Peak so be prepared for wind and cold. Although many campsites are available at Boulder Field, the setting is austere.

NARRATIVE: Turn left at the sign and cross a bridge. Pass a beaver pond on your left and come to a junction of a spur to the Storm Pass Trail. Keep right and climb slightly, soon crossing a small bridge then a larger one. Continue rising beneath conifers and aspen and pass Alberta Falls. Wind up, sometimes traveling close to the narrow, deep rocky gorge formed by Glacier Creek, to the junction of the trail to Loch Vale (See Sky Pond Trail) at 1.0 mile.

Keep left, drop slightly and cross the gorge on a sturdy bridge. The trail rises for a short distance then loses elevation to avoid a band of boulders. Traverse up a wooded slope and at the crest curve right and wind up through stunted timber to the edge of a large basin. Cross it, climbing moderately and enter deeper timber as you approach the eastern wall. Curve left after crossing two streams, between the streams is the Boulder Brook Trail junction. Soon begin a series of long switchbacks. During this stretch you will see weathered remains of the huge Bear Lake fire that devastated this area in 1900.

Climb above timberline on the last switchback and pass between two knobby rock outcroppings, one above the trail and the other downslope. The trail makes a long, sweeping curve along the slope then travels on the level for a few hundred yards to Granite Pass and the junction of the trail to Chasm Lake. Turn right and begin climbing in long, very gradual switchbacks. Come to a bench and walk on the level for a short distance to the north edge of Boulder Field. The trail stops at the rocks and the final several hundred yards to the end of the level stretch is easy boulder hopping.

To reach the Keyhole and the shelter just below it, bear right from the latrine at the south end of Boulder Field and scramble toward the notch on the skyline that has the outline of a keyhole, gaining 400 feet of elevation during the 0.5 mile of travel. The side trip to Chasm View is slightly more strenuous as it involves 0.7 mile and 500 feet of elevation gain. It is reached by bearing left at the latrine and climbing toward the flat place on the skyline just below where the slope to the summit steepens. **USGS:** Longs Peak, McHenrys Peak Quads. **RMNP MAP:** 4.

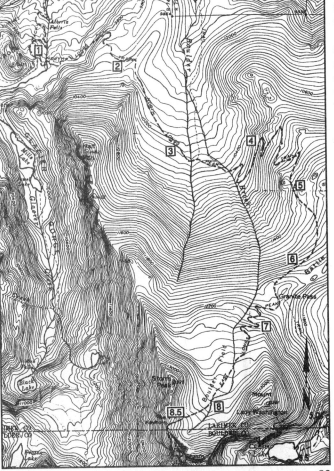

Stone Shelter at Keyhole

Trail Number 21	Trail Name	Map Loc.	Distance	Difficulty	Beginning Elev.	Ending Elev.	Agency
	Eugenia Mine	G 6	1.5 mi	Easy	9,400'	9,908'	RMNP

ELEVATION GAIN: 420 feet. **HIGH POINT:** 9,820 feet. Allow 1 hour one way.

ACCESS: Turn south from U.S. 34 on to U.S. 36 near the east end of Estes Park. Proceed one-half mile then keep right on Colorado 7 and drive 3.3 miles to the junction with Mary's Lake Road. Stay left, continuing on Colorado 7, and drive 5.2 miles to a large sign on your right marking the road to Longs Peak Area. If you are approaching from the south, travel 12 miles north on Colorado 7 from the junction of Colorado 7 and 72. Turn west and drive along the paved road. One mile from the highway turn left at a fork as indicated by the sign pointing to Longs Peak Trail. Drive the short distance to the large parking area at the end of the road.

ATTRACTIONS: A large, dilapidated log cabin and some machinery near the tailings are the most obvious remains of the Eugenia Mine that was active during the first part of the 1900's. The hike is easy and short and you will want to allow extra time to explore the workings.

NARRATIVE: The trail begins just south of the ranger station and climbs at a moderate grade through woods of conifers and aspen. After 0.6 mile come to the junction of the trail to Chasm Lake. Turn right and walk on the level through a grove

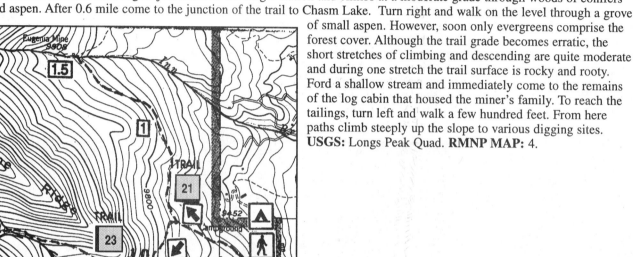

of small aspen. However, soon only evergreens comprise the forest cover. Although the trail grade becomes erratic, the short stretches of climbing and descending are quite moderate and during one stretch the trail surface is rocky and rooty. Ford a shallow stream and immediately come to the remains of the log cabin that housed the miner's family. To reach the tailings, turn left and walk a few hundred feet. From here paths climb steeply up the slope to various digging sites.

USGS: Longs Peak Quad. **RMNP MAP:** 4.

Remnants from the Past

Trail Number 22	Trail Name	Map Loc.	Distance	Difficulty	Beginning Elev.	Ending Elev.	Agency
	Twin Sisters Peaks	G 6	3.5 mi	Mod/Diff	9,111'	11,413'	RMNP

ELEVATION GAIN: 2,375 feet. **HIGH POINT:** 11,413 feet. Allow 2 1/2 to 3 hours one way.

ATTRACTIONS: Before the turn of the century, the peak was known as Lily Mountain from the western side and Twin Sisters from the eastern, probably because the rocky double summits appear as one when seen from certain angles. The view from the lookout on the most westerly (and lower) of the twin peaks extend north past Fort Collins, east over the plains, south to Mount Evans and beyond Longs Peak to near the northwestern boundary of Rocky Mountain National Park. The grade is moderate and steady for almost the entire distance and no water is available at the trailhead or along the hike.

ACCESS: Turn south from U.S. 34 onto U.S. 36 near the east end of Estes Park. Go one-half mile then keep right on Colorado 7 and drive 3.3 miles to the junction with Mary's Lake Road. Keep left, continuing on Colorado 7 and proceed 4.2 miles to a large sign on your left stating Twin Sisters Trail. If you are approaching from the south, travel on Colorado 7 to a point 13 miles north of the junction of Highways 7 and 72. Turn east onto the dirt road and drive for 100 yards. A turnout for parking is just beyond where the road curves left. **NARRATIVE:** Walk up the road 150 feet to where private drives go to the right and left. Take the trail in the middle and after a short, steep climb begin traversing up in woods at a moderate grade. Pass through a gate and continue along the forested slope. Near 1.0 mile begin a series of many switchbacks. Periodically, you will have glimpses of the Indian Peaks area to the southwest and the east face of Longs Peak. As the elevation increases, the trees become shorter, stout and less dense and the forest floor is sprinkled with rocks.

Come to a saddle at 2.5 miles and curve right. Soon the trail becomes steeper and rocky then it switchbacks a few times. Lake Estes and a portion of the plains are visible to the north and just at timberline you can see the lookout above to the southeast. Climb along a boulder field where conies make their homes, switch backing a few times, to a stone shelter at the edge of a narrow saddle between the two summit peaks. Wind up the final few yards to the lookout tower. The Mummy Range, the route of Trail Ridge Road above timberline and Specimen Mountain are on the horizon to the northwest. The mountains beyond Bear Lake and, slightly closer, Wild Basin south of Longs Peak also can be studied. **USGS:** Longs Peak Quad. **RMNP MAP:** 4.

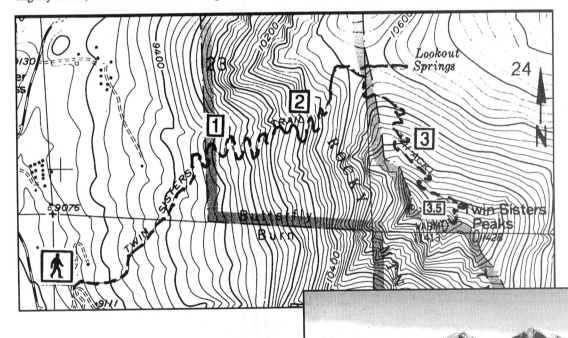

East Face of Longs Peak from Twin Sisters Lookout

Trail Number 23	Trail Name	Map Loc.	Distance	Difficulty	Beginning Elev.	Ending Elev.	Agency
	Chasm Lake	G 6	4.2 mi	Difficult	8,800'	11,800'	RMNP

ELEVATION GAIN: 2,400 feet, loss 100 feet. **HIGH POINT:** 11,800 feet. Allow 3 to 3 1/2 hours one way.
ACCESS: Near the east end of Estes Park turn south from U.S. 34 onto U.S. 36. Proceed one-half mile then keep right on Colorado 7 and drive 3.3 miles to the junction with Mary's Lake road. Stay left, continuing on Colorado 7, and drive 5.2 miles to a large sign on your right marking the road to Longs Peak Area. If you are approaching from the south, travel 12 miles north on Colorado 7 from the junction of Colorado 7 and 72. Turn west along the dirt road that becomes paved at the Park boundary. One mile from the highway turn left at a fork indicated by the sign pointing to Longs Peak. Trail. A large parking area is at the road's end. **ATTRACTIONS:** Probably the best known feature in Rocky Mountain National Park is the sheer east face of Longs Peak. Chasm Lake lies in a cirque at the base of this 2,450 foot wall. Just one-quarter mile below the barren, majestic setting of the lake, a sturdy shelter cabin rests on a grassy little bench where marmots, ground squirrels and conies (rabbit) make their homes.

A side trip to Boulder Field at the 12,800 foot level on Longs Peak can be made by taking the trail from the junction at 4.0 miles and climbing along the northeastern slope of Mount Lady Washington to Granite Pass and the junction of the North Longs Peak Trail. This climb would add a total of three miles and 1,300 feet of elevation gain.

NARRATIVE: The trail begins just south of the ranger station and climbs at a moderate grade beneath a mixture of aspen and conifers. After 0.6 mile come to the junction of the short trail to Eugenia Mine, keep left and soon being switchbacking. Come near Alpine Brook and make one more switchback before traveling in a westerly direction for three-quarters mile. Curve left and come near Alpine Brook for a second time. Climb in a series of short switchbacks then finally cross Alpine Brook on a long foot bridge.

Leave the woods and climb in one long switchback along Mills Moraine to the junction near timberline of the path to Jims Grove. This moraine and a lake and glacier were named for Enos Mills, innkeeper, mountain guide and lecturer who was the individual most responsible for the creation of Rocky Mountain National Park. Turn left and continue gradually up the moraine. Keep on the main trail where you pass a faint path that goes uphill but a short distance further turn right at the well-worn side path marked by a large stone mound. Wind up the open slope at a very moderate grade, following the tall stone mounds. The thin band of the plains, Estes Cone and Twin Sisters Peaks become visible as you climb higher. At the crest come to a four-way junction. The trail to the right is the one you will follow if you make the side trip to Boulder Field. Granite Pass is on the crest of the ridge to the northwest.

Keep straight and drop a few feet, passing a sign indicating the way to Chasm Lake. After a short, moderate stretch of climbing begin a gradual downhill traverse along a steep canyon wall. Travel along the slope above Peacock Pond and Columbine Falls and come to the bench at the head of the valley. The final one-quarter mile to Chasm Lake follows the faint path up the gully behind the shelter. At the crest of the first pitch bear left across grass and rock then turn right and scramble up the final short distance to the rim above the lake. **USGS:** Longs Peak Quad. **RMNP MAP:** 4.

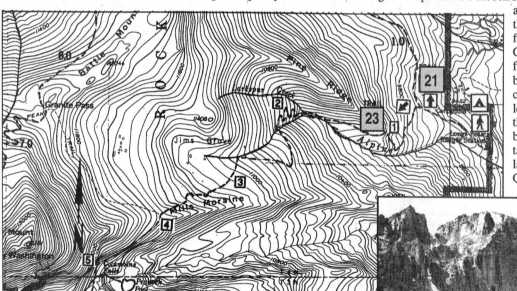

Summit of Longs Peak from Chasm Lake

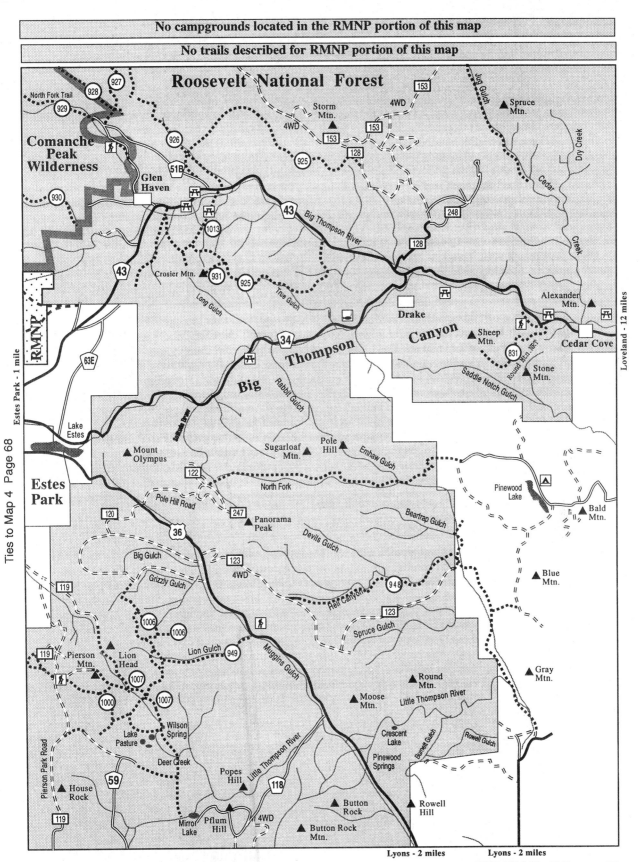

No campgrounds located in the RMNP portion of this map

No trails described for RMNP portion of this map

Roosevelt National Forest

Comanche Peak Wilderness

North Fork Trail

Glen Haven

Storm Mtn.

Spruce Mtn.

Big Thompson River

Crosier Mtn.

Long Gulch

True Gulch

Drake

Canyon

Alexander Mtn.

Sheep Mtn.

Cedar Cove

Thompson

Stone Mtn.

Saddle Notch Gulch

Round Mtn. NRT

Big

Rabbit Gulch

Lake Estes

Mount Olympus

Sugarloaf Mtn.

Pole Hill

Emhaw Gulch

Pinewood Lake

Bald Mtn.

Estes Park

Pole Hill Road

North Fork

Panorama Peak

Devils Gulch

Beartrap Gulch

Big Gulch

Grizzly Gulch

Hall Canyon

Spruce Gulch

Blue Mtn.

Lion Head

Lion Gulch

Muggins Gulch

Pierson Mtn.

Round Mtn.

Gray Mtn.

Wilson Spring

Moose Mtn.

Little Thompson River

Lake Pasture

Deer Creek

Crescent Lake

Pinewood Springs

Burnett Gulch

Rowell Gulch

House Rock

Popes Hill

Little Thompson River

Button Rock

Rowell Hill

Mirror Lake

Pflum Hill

Button Rock Mtn.

Lyons - 2 miles   Lyons - 2 miles

RMNP Map 5

No campgrounds located in the RMNP portion of this map

Ties to Map 3 Page 62

State Hwy. 125 - 12 miles

Estes Park - 35 miles

Bower Gulch Protection Area

Apiatan Mtn.

190

Stillwater Creek

Stillwater Pass Road

123

4

Grand Lake

278

East Inlet Trail

24

Echo Creek

Ranger Cr.

25

Rocky Mountain National Park

Mount Wescott

115

Gold Run

Gold Run

1

Shadow Mtn. Lake

Shadow Mtn. Dam

Pole Creek

Mount Bryant

Arapaho N.R.A.

Trail Mtn.

41

2

Green Ridge

Columbine Creek

3

10

Lake Granby

Indian Peaks Wilderness

Twin Peaks

Walden - 42 miles

602

Rocky Point

10

Twin Creek

3

Willow Creek Reservoir

90

5

Willow Creek

Granby Dam

Granby

125

Inspiration Point

4

125

Hot Sulphur Springs - 7 miles

34

Doe Creek

Strawberry Lake

59

Windy Gap Reservoir

Hankison Reservoir

126

Strawberry Creek

Arapaho National Recreation Area

Lonesome Peak

Granby

60

Caribou Trail

9

Mount Chauncey

Sulphur

Behler Creek

129

Meadow Creek Reservoir

40

55

Eightmile Creek

128

129

Trail Creek

Ninemile Creek

84

Hurd

Hurd Creek

128

Winter Park - 11 miles

Winter Park - 7 miles

2	Trail Number

RMNP Map 6

Trail Number 24	Trail Name	Map Loc.	Distance	Difficulty	Beginning Elev.	Ending Elev.	Agency
	East Inlet	E 7	7.0 mi	Difficult	8,391'	10,200'	RMNP

LAKE VERNA -- ELEVATION GAIN: 1,850 feet. **HIGH POINT:** 10,200 feet. Allow 4 to 4 1/2 hours one way
ACCESS: Proceed on U.S. 40 to the west end of Granby and turn north on U. S. 34. Drive 14 miles to the junction with Colorado 278, keep right, and follow the sign to Grand Lake and Village. After one-third mile the road forks. Keep left, as indicated by the sign pointing to the Big Thompson Irrigation Tunnel and travel 2.4 miles to a sign stating Adams Falls Trailhead Parking. Keep left and after 200 yards enter a large parking area marked by a sign identifying the East Inlet Trail.
ATTRACTIONS: For almost its entire length the trail follows along Grand Lake's East Inlet Creek as it winds through woods, large meadows and valleys, and drops over rock bands. Another trip, (North Inlet), follows North Inlet Creek that also empties into Grand Lake. The route passes Lone Pine Lake at 5.0 miles and cross-country side trips can be made beyond Lake Verna to Spirit and Fourth Lakes. During periods of wet weather or early in the season the ground at the campsites around Lake Verna can be very damp so during these times overnight stays may be more comfortable at Lone Pine Lake. **NARRATIVE:** Walk up the old road bed that heads east from the parking area for several yards then curve south, passing outbuildings, to the register. Climb through woods for 0.3 mile to a sign pointing right to the Adams Falls, only a few hundred feet off the trail to the south. Keep left on the main trail and climb slightly to the boundary of the Park. Walk on the level beside East Inlet Creek then travel in woods near the edge of a meadow. Between 1.0 and 1.5 miles hike above a second, larger meadow. Continue up moderately, at one point passing under blasted rocks, then traverse a steep, wooded slope. Make one short set of switchbacks before walking along the base of a rock band. Climb more steeply then wind up around rocks to the crest where you can look back down over the valley. Drop slightly into a swale of boulders and trees, climb a short distance then descend over slabs toward East Inlet Creek. The trail curves sharply to the left a few yards before the stream and travels on the level through woods. Climb at a generally moderate grade and eventually pass a sign identifying the Mt.Cairns campsite. Wind up and after a more moderate stretch cross a bridge at the base of

a rock wall, switchback steeply a few times then hike at an irregular grade to a bridge over East Inlet Creek. Wind up through woods for 0.8 mile to the south shore of Lone Pine Lake. The main trail curves right, passes an outbuilding on your right and continues near the southeastern end of the lake. Cross a rocky slope of large boulders then begin winding up through woods, crossing several bridges over small streams. At the end of the final span turn sharply right and traverse along a rocky wall. Enter the upper portion of the boulder field you crossed earlier then turn left into a steep-sided little valley. Travel at a slight uphill grade just above East Inlet Creek. Further on, rock slides and log jams have dammed the flow of the stream and formed ponds. Continue through deeper woods then drop briefly to the tip of Lake Verna. The trail continues around the north side to the eastern end of the lake. **USGS:** McHenrys Peak, Shadow Mountain, Isolation Peak Quads. **RMNP MAPS:** 6 & 7.

Gorge on East Inlet Creek

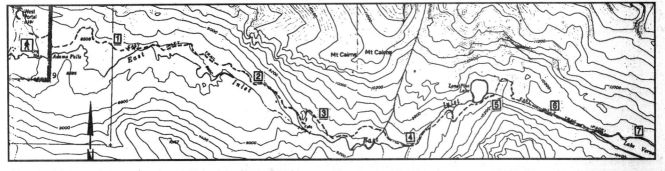

Trail Number **25**	Trail Name	Map Loc.	Distance	Difficulty	Beginning Elev.	Ending Elev.	Agency
	Shadow Mountain Lookout	E 7	4.3 mi	Mod/Diff	8,820'	9,923'	RMNP

ELEVATION GAIN: 1,525 feet. **HIGH POINT:** 9,923 feet. Allow 2 1/2 to 3 hours one way
ACCESS: Drive on U.S. 40 to the west end of Granby and turn north on U. S. 34. Proceed 14 miles to the junction with Colorado 278, keep right and follow the sign to Grand Lake and Village. After one-third mile curve right at a fork, again following the sign to Grand Lake and about 0.5 mile further turn right onto Vine Street where a sign indicates the route to Daven Haven Lodge. After 100 yards turn right then follow the winding main road for 0.5 mile to a junction. Turn left, cross a bridge over a canal and continue 0.25 mile to a sign reading Dead End Road. The last available parking is off the road to the left here. If these few spaces are filled, drive back to another turnoff. **ATTRACTIONS:** From the walkway around the tall stone and wood lookout tower on Shadow Mountain you will be able to see north up the Kawuneeche Valley to the Never Summer Range, southwest to the Gore Range and far to the south over less mountainous terrain. The only source of drinking water along the hike is a creek at 3.4 miles. The lakes visible below the lookout are two of four west side reservoirs for the Big Thompson Project. Water is pumped from Lake Granby up to Shadow Mountain Lake then up to Grand Lake, the only natural lake of the three. From the northeast end of Grand Lake the water flows through the 13.1 mile long Alva B. Adams Tunnel under the southern portion of the Park to the eastern slope. **NARRATIVE:** Walk up the main road (not the one that goes right to private property) for 0.1 mile to a sign stating East Shore Trail and listing several mileages, about 75 feet off the lake side of the road. Climb, then drop for a few yards and travel near the wooded shore of Grand Lake, walking generally on the level. Near 1.2 miles travel away from the lake and come to the junction of the trail to Shadow Mountain Dam. Keep left and after a few yards begin traversing up a wooded slope at a steady moderate grade. About 0.5 mile from the junction switchback to the left and come to a ridge top. Curve left and walk along the narrow crest where you occasionally will be able to see the lookout. Continue climbing through woods and curve right where the ridge you have been following merges with a larger slope. Rise moderately past an area where several small ridges appear downslope. Travel on the level then resume climbing and pass beneath a knob of rocky outcroppings. Walk north, traveling along the slope of a side canyon, then at its head curve right and traverse the opposite wall. At the end of this long, straight stretch come to Sanger Creek and switchback up to the left. Climb in short switchbacks for one-half mile to a saddle. Turn left, walk along the crest for a short distance then wind up a hump to the lookout. A cluster of rocks to the north of the building affords a good place to enjoy your lunch. The actual summit of Shadow Mountain is 0.75 mile to the southeast. Outbuildings are down slope from the hitching rail several yards before the lookout. **USGS:** Shadow Mountain Quad. **RMNP MAP:** 6.

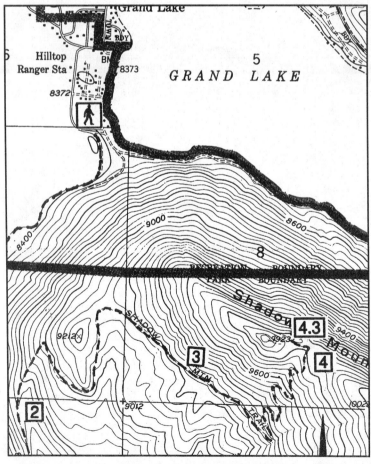

Rock Formations Above Trail

No campgrounds located in the RMNP portion of this map

Ties to Map 4 Page 68

Estes Park - 13 miles

Rocky Mountain National Park

Andrews Peak

Keplinger Lake

Mount Meeker

Dragons Egg Rock

Meeker Park Overflow

Mount Cairns

Lone Pine Lake

24

Lake Verna

Mount Alice

Trio Falls

27

Hunters Creek

Lookout Mtn.

Horsetooth Peak

26

Meeker Park

E. Inlet Trail

Mount Craig

Spirit Lake

Fourth Lake

Thunder Lake Tr.

Sandbeach Lake

Twin Lakes

Wild Basin RS Falls

1

Olive Ridge

Fifth Lake

Thunder Lake

28

Eagle Lake

Allenspark

Paradise Creek

Frigid Lake

Isolation Peak

Mahana Peak

Bluebird Lake Tr.

Ouzel Lake

St. Vrain Creek

7

Mount Adams

Adams Lake

Bluebird Lake

29

Ouzel Peak

Copeland Mtn.

Finch Lake

30

Meadow Mtn.

Willow Creek

116

Watanga Mtn.

Ogalalla Peak

Peak Reservoir

Coney Creek

Rocky Mtn. National Park

St. Vrain Mtn

915

4WD

Rock Creek

Roosevelt National Forest

Watanga Lake

St. Vrain Glacier

St. Vrain Glacier Trail

915

202

203

Roaring Fork

Hiamovi Mtn.

Cooper Peak

917

910

Red Deer Lake

824

910

Middle St. Vrain

Island Lake

Marten Peak

Buchanan Pass

911

4WD

114

2

Mount Irving Hale

Long Lake

Gourd Lake

21

Buchanan Pass

2

Sawtooth Mtn.

914

911

Stapp Lakes

4WD

3

Indian

Coney Lake

Beaver Res.

816

Monarch Lake

Buchanan Creek

1

Thunderbolt Mtn.

Upper Coney Lake

913

Audobon Trail

Mount Audobon

911

Beaver Creek

815

Peaks

Cascade Creek

1

Paiute Peak

Blue Lake

Mitchell Lake

909

Wilderness

6

907

Mt. Toll

912

911

Pawnee Peak

Crater Lake

Lone Eagle

Pawnee Pass

Long Lake

112

Red Rock Lake

Duck Lake

835

6

Creek

Mount Achonee

Shoshoni Peak

Isabelle Glacier

908

Lake Isabelle

907

906

Brainard Lake

4

4WD

Ward

Fair Glacier

Apache Peak

Left Hand Park Reservoir

Niwot Mtn.

72

Meadow Creek Reservoir

Arapaho National Recreation Area

Navajo Peak

Green Lakes

Niwot Ridge

906

93

Arikaree Peak

Kiowa Peak

Island Lake

Silver Lake

50

Mtn. Research Station

298

Glacier Lake

Satanta Peak

North Arapaho Peak

Arapaho Glacier

Goose Lake

905

11

South Arapaho Peak

Caribou

918

5

72

7

Columbine Lake

Lake Dorothy

904

905

Rainbow Lakes

103

Mount Neva

Nederland - 5 miles

2 Trail Number

RMNP Map 7

Ties to Map 6 Page 88

Allenspark - 16 miles

Trail Number 26	Trail Name	Map Loc.	Distance	Difficulty	Beginning Elev.	Ending Elev.	Agency
	Sandbeach Lake	G 7	4.2 mi	Mod/Diff	8,312'	10,295'	RMNP

ELEVATION GAIN: 1,965 feet. High point: 10,295 feet. Allow 2 1/2 hours one way.

ACCESS: Near the east end of Estes Park turn south from U.S. 34 onto U.S. 36. Drive one half mile then keep right on Colorado 7 and proceed 11 miles to Meeker Park. One and one half miles further to a sign for the Wild Basin area. If you are approaching from the south, drive 8.1 miles north on Colorado 7 from the junction of Colorado 7 and 72. Turn north onto the first road and drive past Wild Basin Lodge. The Sandbeach Lake sign identifies the trailhead opposite the Ranger Station. **ATTRACTIONS:** Wild Basin is located in the southeastern corner of Rocky Mountain National Park and the five main trails that penetrate the area are described in this guide. Hike No. 30 goes to Finch Lake, the most southerly trip in the Park, and Trail No.'s 26, 27 and 28 travel up the large, wooded valley formed by North St. Vrain Creek to Lion, Thunder and Bluebird Lakes. The sandy beaches that suggested the name of Sandbeach Lake were inundated when a dam was constructed in the early 1900's.

NARRATIVE: Climb steeply along the wooded southern slope of Copeland Moraine. (The moraine, a falls, lake and mountain in Wild Basin are named for John B. Copeland who homesteaded 320 acres here in the late 1880's.) Where a trail contours to the left keep straight and continue uphill. After 0.2 mile curve west and hike at a much more moderate grade. Enter Rocky Mountain National Park at 0.6 mile and continue traversing above the valley. Make one short set of switchbacks, go over the crest to the other side of the ridge and come to the junction of the Meeker Park Trail.

Keep straight (left) and travel uphill for a short distance. Walk on the level for about 0.2 mile then climb again to the crossing of Campers Creek. The trail turns left on the opposite side of the stream and traverses up to a ridge top. Turn right and walk along the crest then begin a series of minor ups and downs, passing through a grove of aspen and low bushes during one stretch. Continue up through woods and cross Hunters Creek at 3.1 miles. Climb is moderately steep then resumes traveling at a more gentle grade. A short distance beyond a stretch where the trail has been built up above the surrounding terrain come to the northeast end of Sandbeach Lake. **USGS:** Allens Park Quad. **RMNP MAP:** 7.

First Snow of Autumn

27 Trail Number	Trail Name	Map Loc.	Distance	Difficulty	Beginning Elev.	Ending Elev.	Agency
	Lion Lake No 1	G 7	8.0 mi	Difficult	8,723'	11,130'	RMNP

ELEVATION GAIN: 2,630 feet. High point: 11,130 feet. Allow 3 1/2 to 4 hours one way.

ACCESS: Turn south from U.S. 34 onto U.S. 36 near the east end of Estes Park. Drive one half mile then keep right on Colorado 7 and proceed 11 miles to Meeker Park. One and one half miles further to a sign to Wild Basin area. From the south, drive 8.1 miles north on Colorado 7 from the junction of Colorado 7 and 72. Turn onto the Wild Basin Road passing the Ranger Station, go around the east and south shores of Copeland Lake. Continue for two miles to the road's end at a picnic area and the Wild Basin Ranger Station. The trail begins at the southwest side of the parking area near a large bulletin board at the entrance to the turn-around. **ATTRACTIONS:** The hike to Lion Lake No.1 is one of the most scenic trips in Wild Basin. Numerous rock outcroppings are separated by fields of grass and the tarns and clusters of conifers scattered over the terrain add to the already attractive landscape. The numerous benches above the lake invite exploration and if you are backpacking and have time, you are encouraged to take the trails to Thunder Lake and Bluebird Lake. **NARRATIVE:** Immediately cross two large bridges then travel at a very moderate grade with a few short, gentle stretches of ups and downs. After 0.2 mile pass a sign on your left indicating the path to Copeland Falls and continue along the wooded slope. At 1.4 miles come to an elaborate bridge over North St. Vrain Creek. Climb more noticeably through deeper woods for .04 mile to the junction of the spur to Finch Lake Trail at Calypso Cascades. Curve right and cross the complex of streams on a bridge. Although the grade still is irregular, the trail begins climbing more consistently. Three quarters mile beyond the Cascades pass an outbuilding to your right off the trail just before coming to a sign identifying Ouzel Falls.

Cross the stream on a bridge and 200 yards further come to a viewpoint over the valley and across to Longs Peak and Mt. Meeker. The trail drops to avoid a high rock band then resumes rising moderately to the junction of the trail to Bluebird Lake. Keep straight (right) and continue up at a moderate grade. Cross a bridge, curve sharply left at the crest above the span and follow the trail that parallels closest to the stream. Hike at a quite moderate grade then climb more noticeably to the junction of the trail to Thunder Lake.

Turn right and climb very steeply along the rocky trail for 100 yards. The grade then becomes more moderate and the surface smooth as the trail winds up around many boulders. Although generally uphill, the circuitous route also has a few level stretches. Climb to a flat ridge crest and walk on the level before going up over a hump. Descend into deep woods then climb steeply. After the grade moderates come to the end of a grassy swale where the trail becomes faint. Walk through the little valley and where you pass a tarn on your left turn and go around its western side. Drop gradually for several yards and come to the southern tip of Lion Lake #1. An explorative trip would be to climb cross country to Lion Lake No. 2 and Snowbank Lake. **USGS:** Allens Park, Isolation Peak Quads. **RMNP MAP:** 7.

Chiefs Head Peak

	Trail Name	Map Loc.	Distance	Difficulty	Beginning Elev.	Ending Elev.	Agency
28 Trail Number	**Thunder Lake**	G 7	6.8 mi	Difficult	8,723'	10,564'	RMNP

ELEVATION GAIN: 2,170 feet. **HIGH POINT:** 10,650 feet. Allow 3 to 3 1/2 hours one way.

ACCESS: Turn south from U.S. 34 onto U. S. 36 near the east end of Estes Park. Drive one-half mile then keep right on Colorado 7 and proceed 11 miles to Meeker Park. One and one-half miles further to a sign for the Wild Basin Area. From the south, drive 8.1 miles north on Colorado 7 from the junction of Colorado 7 and 72. Turn onto the Wild Basin Road pass the Ranger Station around the east and south shores of Copeland Lake. Continue for two miles to the road's end at a picnic area and a Ranger Station.

ATTRACTIONS: All water in Wild Basin eventually drains into North St. Vrain Creek and the entire hike to Thunder Lake follows beside the stream or high along the slopes of the valley carved by the flow. In the late 1830's one of the St. Vrain brothers and a second man established a store near the present town of Gilcrest The post, where Indians exchanged furs for items such as mirrors, eventually become know as Fort St. Vrain and the stream that flowed into the South Platte River 0.5 miles to the west of the store was called St. Vrain Creek.

If you are backpacking on this hike you can see more of the scenic terrain in the Basin by following the trails to Lion and Bluebird Lakes.

NARRATIVE: The trail begins at the southwest side of the parking area near a large bulletin board you pass on your left as you enter the turnaround. Immediately cross two large bridges then travel at a very moderate grade with a few short, gentle stretches of ups and downs. After 0.2 mile pass a sign on your left indicating the path to Copeland Falls and continue along the wooded slope. At 1.4 miles come to a bridge over North St. Vrain Creek then climb more noticeable through deeper woods for 0.4 mile to the junction of the spur to the Finch Lake Trail at Calypso Cascades.

Curve right and cross the complex of streams on a bridge. Although still irregular, the trail grade begins rising more consistently. Three-quarters mile beyond the Cascades pass an outbuilding to your right off the trail just before coming to Ouzel Falls. Cross the stream on a bridge and 200 yards further come to a viewpoint over the valley and across to Longs Peak and Mt. Meeker. The trail drops to avoid a high rock band then resumes climbing moderately to the junction of the trail to Bluebird Lake. Keep straight (right) and continue up at a moderate grade. Cross a bridge, curve sharply left at the crest above the span and follow the trail that parallels closest to the stream. Hike at a quite moderate grade then climb more noticeably to the junction of the trail to Lion Lake.

Keep left and continue traversing along the slope. Switchback once and pass through an area of stunted trees and rocky terrain. Enter deeper woods and soon have glimpses into the Eagle Basin area high on the slopes across the deep valley. Cross a good-sized stream and continue climbing at a very moderate grade, passing a small meadow on your left at one point. Cross a smaller brook and at the end of one final uphill stretch come to a sign stating Thunder Lake. The trail curves left and winds down to the grassy area at the east end of the lake. An outbuilding is located to the right of the trail just before the clearing. A well-defined trail continues past the patrol cabin and along the north shore. **USGS:** Allens Park, Isolation Peak Quads. **RMNP MAP:** 7.

Thunder Lake 1973

Trail Number **29**	Trail Name	Map Loc.	Distance	Difficulty	Beginning Elev.	Ending Elev.	Agency
	Bluebird Lake	G 7	6.3 mi	Difficult	8,723'	10,950'	RMNP

ELEVATION GAIN: 2,480 feet. **HIGH POINT:** 10,980 feet. Allow 4 to 4 1/2 hours one way.

ACCESS: Turn south from U.S. 34 onto U. S. 36 near the east end of Estes Park. Drive one-half mile then keep right on Colorado 7 and proceed 11 miles to Meeker Park. One and one-half miles further to a sign for the Wild Basin Area. From the south, drive 8.1 miles north on Colorado 7 from the junction of Colorado 7 and 72. Turn onto the Wild Basin Road past the Ranger Station around the east and south shores of Copeland Lake. Continue for two miles to the road's end at a picnic area and a Ranger Station.

ATTRACTIONS: Although not harmonious with the alpine beauty of the terrain surrounding Bluebird Lake, the remains of the massive dam across the lake's outlet are intriguing. The first dam at Bluebird Lake was built in 1902 by a family who operated their ranch on North St. Vrain Creek as a resort. Another, larger dam was constructed from 1912 to 1919 by the second owners of the reservoir.

NARRATIVE: The trail begins near a large bulletin board you pass on your left as you enter the parking area. Immediately cross two large bridges then travel at a generally very moderate grade with a few short, gentle stretches of ups and downs. After 0.2 mile pass a sign on your left indicating the path to Copeland Falls and continue along the mostly wooded slope. At 1.4 miles come to an elaborate bridge over North St. Vrain Creek. Climb more noticeably through deeper woods for 0.4 mile to the junction of the spur to the Finch Lake Trail at Calypso Cascades.

Curve right and cross the complex of streams on a bridge. Although the grade still is irregular, the trail begins climbing more consistently. Three-quarters mile beyond the Cascades pass an outbuilding to your right off the trail just before coming to a sign identifying Ouzel Falls. Cross the stream on a bridge and 200 yards further come to a viewpoint over the valley and across to Longs Peak and Mt. Meeker. The trail drops to avoid a high rock band then resumes climbing moderately to the junction of the trail to Lion and Thunder Lakes.

Turn left and climb steeply for a short distance before curving left and rising more moderately along an old road bed. Come to a ridge top and curve right. Walk along the crest at a very moderate grade and further on begin traversing along the wooded slope. Resume climbing more noticeable and at 4.8 miles come to the junction of the short spur trail to Ouzel Lake. Keep straight (right) on the main route and continue erratically uphill. Drop slightly and traverse above lily pad covered Chickadee Pond. Pass through a small boulder field then resume winding up through woods.

Come to a grassy swath below a rock wall and climb through the open area. Above the steep meadow the trail crosses to the other side of a rocky, little canyon and climbs to the edge of a small boulder field. Cross the rubble then reenter vegetation and walk for a few hundred feet to a sign pointing to Bluebird Lake. Turn left and cross Ouzel Creek on a log. Turn right and after several yards resume hiking on a well defined trail. Wind very steeply up a series of ledges toward the basin that holds Bluebird Lake. The slope covered with spires above the lake to the south is the steep, northwestern face of Copeland Mountain. **USGS:** Allens Park, Isolation Peak Quads. **RMNP MAP:** 7.

Bluebird Lake

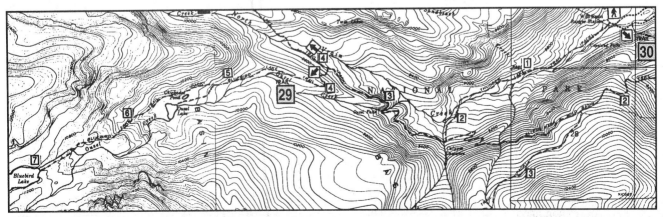

Trail Number 30	Trail Name	Map Loc.	Distance	Difficulty	Beginning Elev.	Ending Elev.	Agency
	Finch Lake	G 7	4.5 mi	Difficult	8,723'	9,912'	RMNP

ELEVATION GAIN: 1,660 feet, loss 220 feet. **HIGH POINT:** 10,160 feet. Allow 2 1/2 to 3 hours one way.

ACCESS: Turn south from US 34 onto US 36 near the east end of Estes Park. Drive one-half mile then keep right on Colorado 7 and proceed 11 miles to Meeker Park. One and one-half miles further to a sign for the Wild Basin Area. From the south, drive 8.1 miles north on Colorado 7 from the junction of Colorado 7 and 72. Turn onto the Wild Basin Road pass the Ranger Station around the east and south shores of Copeland Lake. Continue for almost two miles to a sign on your left indicating trial to Finch and Pear Lakes. Parking spaces for a few cars are available off the road.

ATTRACTIONS: Several lakes in Rocky Mountain National Park have been named for species of birds found here. Examples include Ouzel, Junco, Pipit, Eagle, Bluebird and Ptarmigan Lakes. Probably the birds you are most likely to see at Finch Lake are the resident Clark Nutcrackers (camp robbers) who aggressively seek to share your lunch.

During portions of the hike to Finch Lake you will have views of Twin Sisters and Pagoda Peaks, Mt. Meeker and the southwest side of Longs Peak. The trail continues two miles beyond Finch Lake and climbs 670 feet to Pear Lake. Since the latter is located just below timberline, most of the hike to it is through dense timber.

NARRATIVE: Walk on the level for several yards then curve left and begin a long, steady traverse up the wooded slope of a large lateral moraine. At 0.8 mile come to a clearing at the crest and turn right. Walk through woods of aspen and conifers and after a flat area traverse very gradually downhill to a meadow and the junctions of the trails to Allens Park and Meadow Mountain Ranch.

Keep right on the most northerly trail and walk through a grove of aspen toward the head of the little valley. Reenter the woods and begin climbing. After one set of loose switchbacks traverse along the northern flank of Meadow Mountain several yards uphill from where the slope begins dropping steeply to the valley floor. Walk at a moderate grade then climb more noticeably and come to the junction of trails to Wild Basin Ranger Station and Allens Park. (In 1979 a major forest fire burned this area.) Keep straight and continue climbing. The trial drops slightly then travels along the wooded slope in a series of slight ups and downs. Cross an area of several small streams, the last and largest of which flows throughout the entire summer. After a short distance begin descending and drop for 0.3 mile to the wooded shore of Finch Lake. **USGS:** Allens Park Quad. **RMNP MAP:** 7.

Trail Near Allens Park Junction

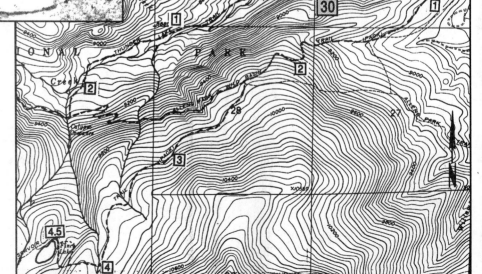